ELECTRO-METALLURGY

Practically Treated

By ALEXANDER WATT, F.R.S.S.A.

LECTURER ON ELECTRO-METALLURGY, ETC.; FORMERLY ONE OF THE EDITORS OF "THE CHEMIST"

NEW EDITION

NEW YORK:

VIRTUE AND YORSTON,

No. 12 DEY STREET.

PREFACE TO THE SECOND EDITION.

It is with much pleasure that the author of this little work again addresses his readers, more especially as he is enabled to give some additional information of a practical nature, which he trusts will be found acceptable.

As the additional matter is given in continuation of the Appendix (commencing at sec. 47, page 113), it is hoped that it will be found more easy for reference than if given in the body of the work.

MERTON ROAD, WANDSWORTH.
March 23, 1868.

CONTENTS.

ELECTRO-METALLURGY,

PRACTICALLY TREATED.

INTRODUCTION.

FROM a simple art almost accidentally discovered, electro-deposition has become a most extensive branch of manufacture in this and other countries. At first "electrotyping" was a very pretty and interesting toy, with which the rising generation of chemical philosophers used to amuse their friends and themselves, as they exhibited the wonderful effects which could be produced by a small galvanic battery, whose poles were immersed in a solution of sulphate of copper. How eagerly the eye used to watch the wondrous deposit of metallic copper upon the plumbagoed impression of some favorite seal! And, how delighted was the youthful philosopher when, on removing the metallic deposit from the sealing-wax impression, he found a faithful representative of the original, as capable, in its turn, of giving a sharp and delicate impression as the original itself!

This great discovery was made in this country by Mr. Spencer, and in Russia, by Professor Jacobi, nearly at the same time; so that the credit of the invention is equally due to each of them. But, as is frequently the case, it devolved upon others to turn the invention to great practical account, and as a matter of course, to a remunerative one also.

So great was the interest felt in this country—nay, almost all over the civilised world, when this beautiful discovery was made known, that persons of every grade in life devoted their attention to it. The student, the mechanic, the artist, the nobleman and the chemist, with equal zeal, though with different views, deposited copper from its solution by electro-chemical agency. Every one had his set of electrotyping apparatus, and his bath of sulphate of copper. Even among the fair sex would be found many a skilful manipulator, and in such hands, how could the art fail to give beautiful results! Everywhere this art was in vogue, and whilst it was being studied as an amusement by some, others were turning their attention to its commercial value, with a view to making it subservient to the useful purposes of life. So that in a very short time this country was well stocked with a new class of competitors—electrotypists, stereotypists, electro-platers, gilders, &c., &c.; and now that the electro-mania has subsided, it has settled itself down into a very comfortable, highly lucrative and legitimate business.

Nearly a quarter of a century having passed away since the introduction of the electro-metallurgical art, it is not to be wondered at that it has ceased to enjoy that popularity which at one time, as we have said, placed it in the hands of all as a fashionable amusement; for those who twenty years ago delighted in the art as an interesting novelty, are now many of them parents of another generation, which may, in its turn, seek to know the uses of the electric current in the deposition of metals upon each other. To aid this class of rising experimentalists, is now our pleasing task.

To those who desire to practise the arts of electro-gilding, plating, &c., with a view to applying the same to commercial purposes, it is hoped that the present work will prove of service, since it is the intention of the author to make it entirely of a practical nature, and as free as possible from technical expressions.

Having been for nearly twenty years practically engaged in the art of electro-plating and gilding, on a very extensive scale, during which period many thousands of ounces of the precious metals have been deposited by me from their various solutions, and having paid great attention to the subject of electro-deposition generally, I have, in common with others, met with many difficulties which careful experiment and perseverance have overcome. Therefore, in laying before my readers the results of my own practical experience, it is with the hope that they may prove useful to those who pursue the study of electro-deposition, either for instructive amusement or profit.

As I have, I believe, been more successful in some of my operations than many of my fellow-labourers, I will carefully describe those processes which I have found to answer best, from their certainty, economy and simplicity, and pass in review the processes usually employed by others, explaining the causes of failure and disappointment so frequently accompanying their adoption.

To render myself as intelligible to the working electro-plater and the amateur, as to the more scientific reader, I will fully explain the meaning of any technical terms which may necessarily occur in the way, so that he may not fall into errors which too frequently—more especially in a chemical art—retard the progress of study.

In depositing metals from their solutions, many forms of galvanic battery are employed. Among those most commonly known are Daniell's, Smee's, Wollaston's and Bunsen's. The first of these, Daniell's battery, has been almost abandoned, owing to the trouble which it involves to keep it in good working order. The second, Smee's battery, although far from economical, and somewhat uncertain in its action, is still employed by some, owing to the great *intensity* of the current which it produces (a quality of but little service to the electro-plater when the *quantity* is deficient, as we will presently explain). The third, Wollaston's battery, by far superior to the latter for electro-metallurgical purposes, as it yields a great *quantity* of electricity of considerable tension, is also frequently employed, or rather modifications of the same arrangement, which are fitted up with but little trouble and expense; whilst Bunsen's battery is only capable of being employed in depositing those metals which require a current of great intensity, as well as quantity. This battery, however, is quite unfit for the purpose of depositing either gold, silver, or copper.

It must be borne in mind, that in order to ensure a perfectly smooth, equal and *reguline* deposit on a metallic surface, the battery to be employed should yield a *considerable quantity of electricity of sufficient intensity* to work with activity and *uniformity*. A battery constructed with a large surface of positive and negative elements—as zinc and copper for instance—will yield a current of such feeble intensity in proportion to that quantity, that, when employed for the purposes of electro-deposition, the deposit takes place very slowly; whilst a battery consisting of a great number of small plates or cells, alternately arranged, would not only deposit the metal in a granular or pulverulent form, but would actually decompose the solution itself. Consequently, in order to obtain a good reguline deposit of any metal, a battery should be employed whose positive and negative elements are in such relative proportion as to yield a current of quantity electricity possessing sufficient intensity to enable that quantity to work well.

A form of battery which I have found most constant and certain in its action, I will describe further on, as also one which is much used in extensive operations where great power is required to deposit large quantities of metal, as in the processes of electrotyping and electro-plating.

Faraday employs the terms *anode, anelectrode,* or *positive electrode,* for the positive pole of the battery—*i.e.,* the wire which proceeds from the copper element in a battery; and *cathode, cathelectrode,* or *negative electrode,* for the negative pole—that which proceeds from the zinc element. Professor Daniell, however, objecting to the terms *anode* and *cathode,* proposed the adoption of *zincode* and *platinode,* to distinguish the positive and negative poles; but as the elements of a battery are not necessarily composed of zinc or platinum, and as, independently of the great weight which must always attach to any system propounded by Mr. Faraday, it would sound rather unmusical to speak of *leadodes, arbono des,* or *copperodes,* when describing the poles of a battery with an element of lead, carbon, or copper, I prefer adopting Faraday's nomenclature.

The electricity generated in a cell passes from the zinc to the copper element of the battery, and from thence it proceeds along the wire issuing from the copper, traverses the solution, and returns to the cell through the wire which is attached to the zinc element, and so on. The zinc is the *positive* and the copper the *negative* element, but the end of the wire attached to zinc becomes the *negative pole,* whilst that proceeding from the copper becomes the *positive pole.*

The *anode,* or positive pole, is that wire which is attached to the copper cylinder or plate of a battery; and to this wire or pole is suspended, in close contact, the sheet or plate of metal which is destined to re-supply the solution with the amount of metal which it loses by the deposition which takes place on the cathode or article to be coated.

The *cathode,* or negative pole, is the wire which issues from the zinc plate or bar of a battery, and it is this wire or pole, or any metallic surface which may be attached to it, which receives the deposit in the bath.

Professor Faraday denominates the solution, whether it be of silver, gold, copper, or any other metal from which a deposit is to be obtained, the *electrolyte.*

Quantity electricity, as I have already observed, is that kind of current which is produced when the battery is formed of large surfaces of the metallic element; it is this species of electricity which is most useful for the purposes of electro-deposition.

"Experience proves that, in general, the adherence of the oxides and of the metals gold, silver, copper, and lead on metals, is greater as the intensity of the current is less, within certain well-known limits; and as the solution is less concentrated." *

Intensity may be given to the quantity already existing in a series of cells or plates, by increasing their number; thus, by attaching the wire proceeding from the positive pole of one cell to the negative pole of another, and so on, until a compound battery is formed of alternate pairs. A battery thus constructed is well adapted to the purposes of electro-chemical decomposition, or *electrolisation,* the electric light, the giving of shocks, and other powerful effects of electricity; but, unless carefully applied, it would be highly injurious if devoted to electro-metallurgical operations.

An intensity current seldom lasts longer than a few hours, unless fresh exciting fluids be applied to the elements with which it is produced; but a quantity current may continue to be developed from a constant battery for months. I have known a constant battery continue in action for twelve months *without any addition whatever,* at the end of which period it still gave considerable evidences of electrical action.

The Battery.—The battery which I would recommend to the attention of the electro-gilder, and those who desire to deposit metals by electricity on a moderate scale, consists of a cylindrical stone jar A (fig. 1), capable of holding about four gallons; inside this jar is fitted a cylinder of sheet copper C (this may be $\frac{1}{64}$th of an inch

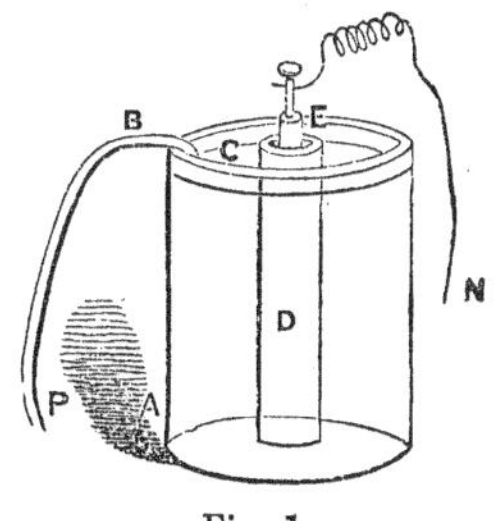

Fig. 1.

* Becquerel, "The Chemist," 1843, vol. iv. p. 400.

in thickness). A strip of the copper cylinder B, about half an inch broad, is cut off to within one inch, so as to form the positive electrode; my motive in doing this is to insure a perfect connection between the positive pole and the cylinder, and to save the trouble of soldering.

A circular piece of wood forms a covering to the jar; in the centre of this cover, a hole about two inches in diameter is bored, to which an ox-gullet D, or weazand, is fastened, extending to the bottom of the jar, the lower end of which is carefully tied with a piece of thick twine; or a porous cell may be used instead if preferred. A zinc bar E (fig. 2) is cast, with a long and tolerably thick copper wire in it, one end of which has been previously coiled into a helix, so as to form a spring, to prevent the breaking off of the wire at its junction with the zinc bar. The ox-gullet, or cell, is now nearly filled with a concentrated solution of common salt, to which a few drops of hydrochloric acid have been added, and the zinc bar immersed in it, but not allowed to touch the bottom of the gullet, or cell, which it may be prevented from doing by attaching a piece of wood across the zinc bar, to suspend it from the cover of the battery. The jar is nearly filled with water acidulated with two pounds of sulphuric acid and one ounce of nitric acid, and the battery is ready for use. P and N (fig. 1) signify positive and negative poles.

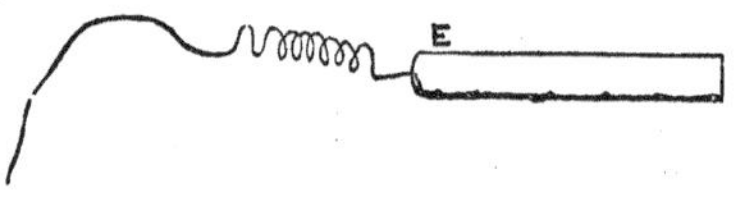

Fig. 2.

In the above form of battery several advantages present themselves; its action is constant, there is but little local action, and consequently but little waste;

its current is regular, and it is very economical in its construction and inexpensive in use.

A compound battery thus constructed will give most powerful effects when a number of cells are used, and it will continue to give these effects for a greater length of time than any battery with which I am acquainted.

In a single cell of this battery, a considerable quantity of electricity is disengaged, of sufficient intensity for small operations, such as gilding and so forth. When it is desired to deposit a large quantity of metal in a given time, several of these cells alternated, that is, having the zinc wire of one cell united to the copper cylinder of the next, and so on, may be employed, by which arrangement a vast amount of metal may be deposited in a short time, when the solution is in good working condition. But it is preferable to unite all the copper wires and the zinc wires, by which arrangement the intensity is not increased.

In working with a Smee's battery in the large way, the rapid consumption of the zinc plates, the furious local action and offensive evolution of hydrogen gas which it is susceptible of, and the trouble and expense of amalgamating the plates, are among the many disadvantages which this battery exhibits to the practical electro-metallurgist; added to which, the current which proceeds from it is far too intense and fluctuating to enable us to obtain a smooth and reguline deposit. But for many experimental purposes this is one of the most convenient and ingenious batteries known, and Mr. Smee deserves the highest credit for its introduction, as its great popularity will testify.

Wollaston's battery, were it not for the trouble and

difficulty of replacing the zinc plates when they are consumed, and the constant application of exciting material which it requires, would be admirably suited to electro-metallurgical operations. A useful modification of Wollaston's battery, however, is now much in use. It consists of a cylindrical stone jar, A, capable of holding about ten gallons; two pieces of sheet copper are fixed upon a wooden support, D. A plate of amalgamated zinc, *c*, is placed in a groove cut in the wooden bar or support between the copper plates. A binding screw is soldered to the copper plates, B *b*, which are united by strips of copper, soldered to them, and a binding screw is to be fastened to the zinc plate. The jar is to be filled with sulphuric acid one part, water fifteen parts. The zinc must be well amalgamated.

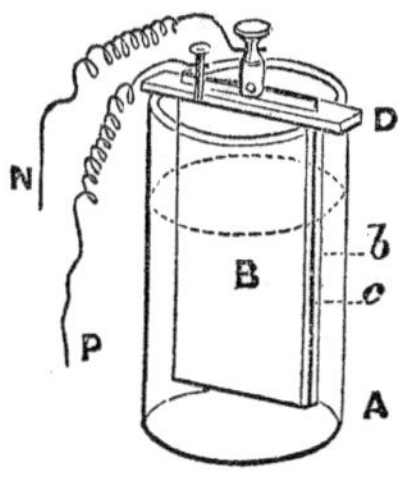

Some electro-platers have employed magneto-electricity for the deposition of metals, but not, I think, with much success; owing, no doubt, to the fact that a current of magneto-electricity would be liable to interruption, or a want of continuity. The action of revolving armatures must necessarily be interrupted, owing to the making and breaking of contact. I maintain that, in order to obtain a fair deposit, the current of electricity applied *must be continuous.*

My brother, Mr. Charles Watt, patented a thermo-electrical battery which, if employed on a large scale, would probably eclipse all other batteries for electro-metallurgical purposes, as it possesses all the advantages of constancy, uniformity, and economy; being, in fact, almost costless in its action.

There are other circumstances besides the power of the battery which affect the nature of the deposit, or the speed with which it is obtained. The solution, or *electrolyte,* may be what is termed a *good* or a *bad* conductor, according to the amount of metal or the proportion of the solvent existing in it; or the extent of surface of anode or positive electrode immersed in the solution while deposition is taking place. If the solution be poor in metal, &c., and the surface of anode exposed to the article which is to receive the deposit be smaller than is required, the operation will go on slowly; whilst, on the other hand, a superabundance of metal and the solvent being in the solution, and the surface of anode exposed being considerable, the deposit may take place so rapidly that it will be thrown off the cathode, or article coated, in the form of a powder, or myriads of minute granules.

Again, the speed with which the deposit is obtained depends upon the temperature of the solution. When the solution is raised to the temperature of 60° C. (140° F.), deposition takes place very rapidly; indeed, in order to bring the solution to a strength which will enable you to use it hot without fear of granular deposition and other imperfections, nearly 75 per cent. of water must be added to it, and the surface of anode immersed be diminished.

In excessively cold weather, I have frequently found a silver solution covered with ice of considerable thickness, and consequently the deposition has taken place more slowly than was desired. In this state the deposit was much harder, and less inclined to be "rough," than when the solution was of a higher temperature. I would at all times prefer working the silver solution at as low a temperature as possible, as I think the deposit, under such circumstances, is in many respects of a superior quality.

Motion will also materially affect electro-deposition. If the solution be too strong; the surface of the anode exposed be excessive; the solution be of too high a temperature; the battery too powerful, or if any one of these circumstances give rise to a pulverulent or granular deposit, or cause the metal to "strip," or peel off the article on which it is deposited, by keeping the negative electrode and the article attached to it in constant and rapid motion until the required coating is obtained, a perfectly smooth, uniform, and tenacious deposit will be secured, though the circumstances referred to be ever so unfavourable. For example, if you attach an article to the negative electrode, and place it in the gilding bath, and if, after a few seconds, you observe that the gold is deposited of a dull brown colour, by very briskly agitating the article in the solution it will instantly become bright and of a good fine-gold colour.

There are circumstances under which no deposition whatever will take place. The following occurrence will illustrate a curious phenomenon which occurred to my brother and myself some years ago. We had been plating large quantities of spoons and forks in an apartment for several years, during which time our operations had been most highly successful, and we had been much praised for the quality of our deposit. One day my brother found, to his great annoyance, that no deposit whatever would take place on any article immersed in the solution. Something was wrong. Entirely new batteries were applied, but with no better success;

fresh solutions were made, but still no deposition of silver took place. The batteries and solutions were next insulated from contact with the ground, as we thought it probable the current was being conducted away somehow or other, and yet no favourable change occurred. Thus matters went on for nearly a fortnight; all hands were idle; the workpeople enjoyed a kind of extended Easter holiday, or were hoping something favourable would "turn up" from day to day. At last, having tried every expedient that suggested itself to our almost distracted senses, it occurred to me that if the solutions and batteries were removed to *another apartment* we might meet with better success. The experiment was tried and it succeeded. Once more we could observe the beautiful deposit of silver upon the metallic surfaces, and all went on well.

Whatever may have been the cause of this inaction, some time afterwards the operations were carried on in the same apartment with perfect facility.

In practising the art of electro-deposition, it is necessary to observe the strictest cleanliness, and to be careful not to allow the solutions in any way to be mixed with each other.

It will be necessary to have various kinds of solutions, of certain strengths, in order to deposit one metal upon another with tenacity and firmness. The same solution will not do well for all metals. It is the neglect of this fact which causes many failures, and many solutions to be spoilt. A solution which will allow a good deposit of silver to take place on copper or brass, will not be applicable to steel, as the silver would instantly blister or peel off the latter. Again, a solution which would deposit a faultless coating of copper on iron would deposit a very bad coating on zinc.

To those who are unacquainted with science, I may observe that they need not be deterred from the study of these arts by any apparent abstruseness which may at first sight, surround it. In the present portion of this work I have been under the necessity of entering chiefly into scientific considerations: but will now commence the details of the various processes of electro-deposition, which I will endeavour to render as simple as possible, in order that they may be fully understood, even by those who now enter upon the study of this subject for the first time.

Electro-deposition of Copper.—Many valuable improvements and additions have been made by the various manipulators in the beautiful art of electrotyping; one of the first of which was Mr. Murray's application of plumbago (carburet of iron), as a coating for surfaces which were non-conductors of electricity.

Electrotypes were originally produced in a cell which formed at the same time the battery and the decomposition bath, thus:—A jar A was charged with a concentrated solution of sulphate of copper ("blue stone" or "blue vitriol"). A porous cell B, a bladder, or a glass tube having one end covered with a piece of bladder, was placed in this solution, and a piece of zinc with a copper wire C attached was placed in this cell, which was then filled with dilute sulphuric acid or salt and water; the object to be copied, being previously prepared, was suspended to the end of this wire D and immersed in the

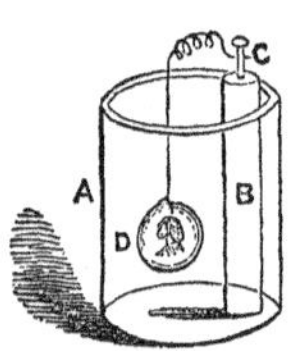

copper solution. This was termed the "single cell" arrangement; it is even now occasionally used by electro-metallurgists in some of their operations.

Subsequently, experimentalists applied a separate battery for the purpose of depositing copper from its solution, and it was found that operations on a large scale could thus be carried on with considerable speed and other advantages. Mr. Mason has the credit of being the first who applied a separate battery to the production of electrotypes.

When a separate battery is used, it is necessary to attach the mould to be copied to the negative electrode,—the wire proceeding from the zinc of the battery, and a piece of sheet copper is attached to the positive electrode—the wire issuing from the copper of the battery. In this arrangement the object to receive the deposit constitutes the *cathode,* and the copper plate the *anode.*

Copper Solutions.—The solution for electrotyping by means of the "single cell" arrangement should be composed of a nearly saturated solution of sulphate of copper, with two ounces of concentrated sulphuric acid added to the gallon of saturated solution; one drachm of arsenious acid (white oxide of arsenic) may be also added to improve the character of the deposit, but this is not indispensable. A little chloride of tin may be substituted for the arsenic.

The sulphate of copper may be dissolved in boiling distilled or rain water, or even common water, and allowed to cool, the sulphuric acid being added when the solution is quite cold.

Sulphate of copper is frequently adulterated with sulphate of iron ("copperas" or "green vitriol"), therefore it is necessary to obtain the article at a respectable establishment; in fact it is advisable always to procure substances required for experiment, or even for more extensive operations, where their purity can be depended upon. If every one adopted this principle, those who vend impure materials would soon be compelled to follow the example of their more honest competitors, and to sell pure articles, however little in accordance with their wishes.

The solution required for depositing copper with a separate battery is composed of—

Sulphate of copper		1 pound.
Sulphuric acid		1 ,,
Water	(about)	1 gallon.

to which may be added a small quantity of arsenious acid or chloride of tin.

PREPARATION OF MOULDS.

The material of which a mould is composed will depend upon the nature of the model; the same composition will not do well for all purposes.

Moulds from Plaster of Paris Models—may be obtained by any of the following methods:—If the object to be copied be a plaster medallion, for instance, let it be placed in a plate or large saucer, with its face upwards, and pour boiling water all round it until it nearly reaches the upper edge of the cast; allow it to remain in the water until the face of the object assumes a moist, but not wet, appearance; then remove it from the plate and surround it with a rim of card or thick drawing-paper, allowing sufficient depth in the rim to hold a requisite quantity of the moulding material. This rim of card may be conveniently kept in its position by sealing-wax. A rim of sheet tin or brass will be found also to answer the purpose very well, but it must be secured to the medallion by means of fine binding wire or a split ring. The medallion must not stand for longer than two or three minutes after it is taken out of the water, before the composition is poured on. It is better to put the rim of card round the cast before immersing it in hot water. The following composition, being melted and at the point of cooling, is then poured into the mould.

White wax	6 ounces.
Spermaceti	1 ,,
Stearine	8 ,,
Carbonate of lead	1 ,,
	16

These compounds should be well melted together, the carbonate of lead being added last, and thoroughly stirred; care must be taken that the heat applied be not sufficient to form air-bubbles. As soon as the composition is poured on the medallion, it is advantageous to quickly stir it with a camel-hair pencil to dissipate any air-bubbles which may have resulted from pouring in the composition too suddenly. Also, the mould thus formed should remain for several hours to become quite cold; the more gradually it cools the better. The rim may now be removed and the mould separated from the medallion. Should there be a tendency for the two surfaces to adhere, the plaster cast may be again placed in boiling water for an instant, when it will come away readily. Sometimes, however, the composition will adhere to the plaster in spite of all precautions, in which case it is advisable to force it asunder, taking care not to injure the composition mould. If some of the plaster is found to adhere to the mould, place the latter in luke-warm water for a short time; this will somewhat soften the adherent plaster, and will enable portions of it to be picked off the surface of the mould, and with a very soft brush much more will come away. Should any plaster still obstinately remain adherent, dry the mould and apply with a thin piece of wood a little sulphuric acid to the fragments of plaster remaining, and leave the mould exposed to the air for some time, when the acid will have attracted a certain quantity of moisture from the air, and their united action will cause the gradual dislodgment of the plaster, which may be brushed away with a soft brush and water.

Gutta percha is another excellent substance for making moulds from plaster of Paris models. The gutta percha must be boiled in water for some length of time until it is quite soft. The object to be copied, if a plaster medallion, should have its surface slightly oiled, and then be provided with a rim as before described, and the softened gutta percha, being wiped dry and rolled into the form of a ball, placed in the centre of the model and worked with the hand until every part of the medallion is covered with it, when a smooth piece of wood (previously greased) may be placed over it and pressure applied until the mould is thoroughly set. In about an hour or so it may be removed from the model. It is necessary to bind the rim round the plaster cast *very tightly*, in order to render the object less liable to fracture and to keep the parts well together if an accident does happen; or the plaster cast may be imbedded in a little melted wax, poured on a plate, previous to the gutta percha being applied; by this means the plaster will be quite secured from fracture. Pressure may be conveniently applied by placing the mould, &c., between two pieces of perfectly flat wood and then screwing them in a vice, taking care that they be properly adjusted so that the pressure may be uniform, or a weight may be placed on the mould, and allowed to remain for half an hour or so.

Moulds in fusible metal may also be obtained from plaster casts. The plaster model should first be well soaked in boiled linseed oil, to which a little "patent dryers" has been added, and allowed to remain for several days before taking the mould, when it will have become exceedingly hard. The mould may then be taken from the plaster cast in the same way as from medals, described further on.

Elastic moulds, as they are termed, may be made from casts in plaster. The composition for this purpose is—

Glue	12 ounces.
Treacle	3 ,,

Soak the glue in sufficient water to render it quite soft. As soon as the glue is quite liquid, add the treacle and mix them well together. The plastic cast must be thoroughly saturated with boiled linseed oil, containing a little "patent dryers," and be laid aside for a day or two, if convenient, to harden before the elastic mould is made from it. This material for moulds is generally applied to objects which are either much "undercut," or are in considerable relief, and from which, consequently, it may be impossible to obtain a perfect copy without this composition is resorted to. The elastic moulds are thus made. If we desire to copy a figure of plaster, after it has been subjected to the linseed oil, &c., let the hollow in the figure be filled up with sand,

and the orifice at its base be well closed with a piece ot card or oilskin pasted over it. The figure is now placed perpendicularly in a jar of cylindrical form, and rather deeper than the height of the bust; the jar should be previously well greased. The plaster cast must have an abundance of oil brushed or poured over it before it is placed in the jar, and the composition is poured in until it covers the bust and is an inch or two above it.

After allowing the mould thus formed to remain for a day or so to become thoroughly set, the jar may be turned upside down, and the mould will readily slip out. A very sharp, bright, and thin-bladed knife, is now passed from the top to the bottom of the figure at its back, very cautiously, and the mould may be opened and the plaster model withdrawn. As soon as the model is removed, the mould, being elastic, will close itself. A strip of oiled paper or rag is now carefully wrapped round the mould, in order that it may retain its proper position: it is a good plan, also, to place three or four pieces of wood of equal thicknesses, at equal distances round the mould, secured by a piece of twine; this will protect the mould from injury. The mould being inverted, is now filled with a mixture of about equal parts of bees'-wax and resin, and a small quantity of plumbago and tallow. The mixture should not be poured in until it is beginning to cool. The whole should be allowed to rest for a few hours until quite cold, when the wooden props and bandages may be removed, the mould reopened, and the composition figure gently withdrawn. The mould will do for future occasions.

When the mould is made of the wax composition, it

should be treated in the following manner. Bend a piece of stout copper wire in such a way that it may, when slightly heated, be conveniently placed round a portion of the edge of the medallion, to which it will adhere firmly when cold. Then apply, with a soft camel-hair, badger-hair, or other very soft brush, finely powdered plumbago (common blacklead will do) until the whole surface of the mould has acquired a metallic lustre. The brush with which the plumbago is applied should be worked *in circles,* so that every little crevice in the mould may be thoroughly coated; it may be advisable also to plumbago the finger and rub the flat surfaces of the mould with it, in order that they may be uniformly blackleaded.

It is sometimes advantageous to breathe upon the surface of the mould when applying the plumbago; care must be taken that the end of the conducting wire attached to the mould, and that part of the composition near it, receive a good coating of the plumbago to insure a perfect connection between the wire and the plumbagoed surface. The edge of the mould should be scraped round with a knife, in order to remove any superfluous plumbago which may have been communicated by the fingers, or otherwise this part of the mould will receive the deposit, and render it difficult to separate the electrotype from the mould. But care must be taken not to remove the plumbago from the wire and adjacent composition.

The mould is now ready to be placed in the solution bath; if it is desired to obtain a good thick deposit, it may be left in the bath for two or three days or even longer. When the mould has received the required coating, remove it from the bath, detach it from tne zinc element, and then gradually loosen the electrotype with the point of a penknife. Should there be any copper deposited on the outer edge of the mould, thereby rendering it difficult to separate the one from the other, this may be broken away and the obstacle thus removed. It is advisable, before taking the electrotype from the mould, to cut off the conducting wire as close to the copy as possible, in order to render the detachment more manageable.

As soon as the electrotype is free, it may be heated to cherry redness in a clear fire, or, which is better, by a blast from a blow-pipe, and when thus *annealed* it will be exceedingly tough, and less liable to be broken. When cool, the electrotype may be plunged into cold water acidulated with sulphuric acid, and allowed to remain in it for some minutes, when it may be rinsed and dried, the edges clipped with a pair of jewellers' shears, and filed to the proper form.

The electrotype may now be polished with rotten stone and oil, and applied with a rather stiff brush. It may then be washed with boiling water and soap, dried, and, lastly, polished with moistened rouge and a soft brush, the plain surfaces being polished with the second finger and rouge.

Previously to polishing the electrotype, the hollow surface at the back may be filled up with pewter solder and lead, thus :—Dissolve a piece of zinc in hydrochloric acid (muriatic acid) and apply a little of the solution all over the back of the electrotype; cut up some pewter solder into small pieces and place them on the back, put the copy on a piece of charcoal, and apply the blow-pipe flame until the solder has "run" into every crevice. Some pieces of lead may now be treated in a similar way to give additional substance to the electrotype, and it is cheaper than solder. The copy may now be bronzed, plated, or gilt, and mounted on a piece of black velvet, or otherwise disposed of, according to the taste of the electrotypist.

Moulds from Metallic Substances may be obtained by any of the following processes:—Suppose it be a medal which we desire to copy, let a stout piece of copper wire be soldered to the edge or back of the medal, or let a thinner piece of wire be twisted tightly round its edge. Then place the medal, face upwards, in a plate containing a little melted wax, suffering the wax to reach about half way up the edge of the medal, then remove it for a moment, and replace in the wax once more to give an additional coating. Or soften a piece of gutta percha, roll it into a ball, and, having cut a hole of the size of the medal in several pieces of card, or one thick piece of cardboard, place the medal, face downwards, between these holes and press the gutta percha on the back of the medal, and put a weight upon it. It may be advisable to coat the back of the medal with a solution of gutta percha, in order to give the lump applied a greater inclination to adhere, or the medal may be somewhat heated before the gutta percha is applied.

The face of the medal is now to be slightly greased either with olive oil, trotter oil, or melted goose fat. This is best done with a camel-hair pencil or a piece of cotton wool. The superfluous oil is then to be removed from the medal by means of a piece of clean cotton wool or a silk handkerchief. Solutions of wax in alcohol or turpentine have been substituted for oil or grease. The surface of the medal may also be plumbagoed with advantage, in which case the oil may be dispensed with. The medal is now to be put into the solution bath, and allowed to remain until sufficiently well coated, when it may be removed, washed, and the mould carefully separated from it.

The mould may now, in its turn, be oiled or plumbagoed, and placed in the bath, and a deposit being allowed to take place upon it, the operator will have obtained an exact representative of the original. This may now be treated in the same way as the electrotype from a wax mould.

The next process for obtaining moulds from metals, consists in first oiling or plumbagoing the surface of the medal, then placing a rim of card round its edge, secured by sealing-wax. Some very fine plaster of Paris is now mixed to the consistence of thick cream, and this is carefully poured over the face of the medal with a table-spoon; a camel-hair pencil is now used to stir the plaster on the medal, in order to dissipate any air bubbles which may have been formed when pouring on the plaster. The brush is quickly plunged into water, and the plaster allowed to remain for an hour or so to harden. When the mould is separated from the medal, it should be placed aside to dry as much as possible, and it must be well charged with melted wax before being plumbagoed. A wire may be firmly twisted round it, and the *connection* between the wire and the mould be secured by brushing the plumbago at that part only where the wire is twisted; otherwise, should the whole of the coil of wire be plumbagoed, there may be considerable difficulty in detaching the copy when the deposit is obtained. As before, all superfluous plumbago should be scraped off the edge of the mould before immersing in the bath.

Gutta percha moulds may also be obtained from metallic substances in the same way as from plaster models.

Sealing-wax has also been employed to obtain moulds from metallic surfaces, but it is not so suitable as either gutta percha or the following :—

After a medal has been oiled or plumbagoed as before, and a rim of card bound round its edge, a mould may be made of the wax and stearine composition, which is melted gradually, and when it begins to solidify, it is carefully poured on the surface of the medal, this being held at a slight angle at the time in order to prevent the formation of air bubbles. If the composition is too hot, or if the mould be too quickly removed from the medal, it will surely adhere. The mould should not be removed for several hours. If, however, with all precaution, the mould has an inclination to adhere to the medal, place them for an instant in hot water to expand the medal, when it will separate easily.

Moulds in *fusible metal*, prepared by various processes, are also obtained from medals, &c. The fusible alloy may be formed from any of the following formulæ :—

Melt together in a crucible or clean ladle

Bismuth	8 ounces.
Lead	5 ,,
Tin	4 ,,
Antimony	1 ,,
	18 ,,

While these substances are being fused, nearly fill a cylindrical jar of considerable depth, with cold water. Cut some hay or straw into pieces of about three inches in length, and place them in the water. Let some person keep this well stirred until the metal is ready to pour. The stirrer is then withdrawn quickly and the melted alloy poured in. This will finely granulate the alloy. The water being now poured off the granulated metal, it may be dried and remelted. By this means the alloy becomes thoroughly well mixed.

Or either of the following mixtures may be treated in the same way.

I.

Bismuth	8 ounces.
Lead	4 ,,
Tin	4 ,,
	16 ,,

II.

Bismuth	9 ounces.
Tin	3 ,,
Lead	6 ,,
	18 ,,

III.

Bismuth	8 ounces.
Tin	3 ,,
Lead	5 ,,
	16 ,,

When a medal is to be copied by the fusible alloy, it should be placed on a smooth piece of wood, and the edge of the medal traced round with a pencil upon the wood; a hole is now to be cut in the wood, as deep as half the thickness of the edge of the medal, and when this is done the medal is to be placed in this cavity, and made fast to it by means of moist blotting-paper or otherwise. (*See engraving.*) Or the back of the medal may be imbedded in a thick paste of plaster of Paris, up to half the thickness of its edge, and the plaster worked up so as to form a kind of handle for the medal, which should not be greased for this purpose. When the medal is secured by either of the above means, a wooden block is to be obtained, a part of which is to be greased a little, and a quantity of the fused alloy poured quickly on it, and this is to be worked up with a thin piece of wood, or card, into a mass of pasty consistence. If a pellicle appears on the surface it must be quickly removed with the card, and the medal be brought suddenly upon the cooling alloy, where it must be held steadily for a few moments until the alloy has quite *set*.

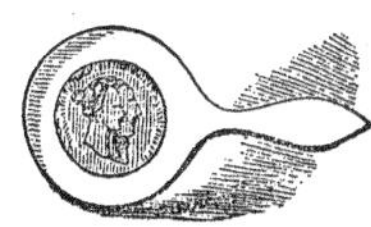

It is absolutely necessary to act with promptness and expertness, in order to obtain good moulds by means of the fusible alloy.

Moulds from Animal Substances.—Let us presume that the object to be copied is a fish. A quantity of plaster of Paris is mixed into a thickish paste, and poured quickly on a piece of plate-glass or sheet tin, slightly greased, to prevent the adhesion of the plaster; or a sheet of paper, greased on one side, placed on a level surface of wood, will answer this purpose very well. The fish may then be laid on its side upon the plaster, and a little gentle pressure applied until one-half of the fish is imbedded. (*See*

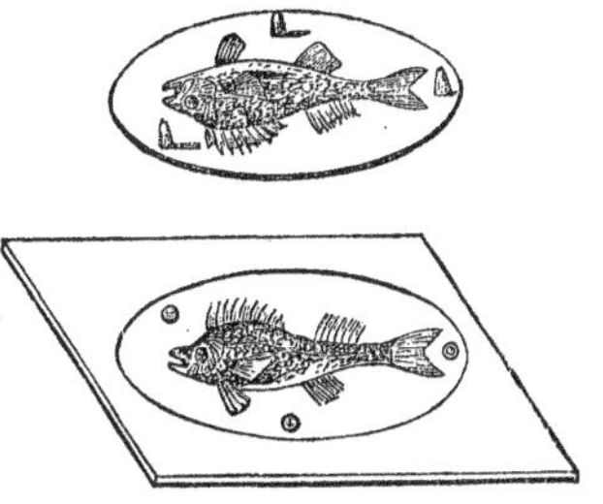

woodcut.) It is advantageous, sometimes, to brush oil over the fish, previous to placing it in the preparation of plaster. As soon as the adjustment of the fish is complete, it may be allowed to remain until the plaster is thoroughly *set*, but not hard; the fish may then be carefully removed from the mould thus formed, and any "ragged edges" which may appear on the mould, may be smoothed with a penknife. Three holes, of a conical form, and at least half an inch deep, should now be bored in the face of the mould, thus,—one near the middle of the fish's back, another below the head, and a third beneath the tail.

The mould must then be brushed over with soap and water, a very soft brush being applied, and the fish is then carefully replaced in its former position. Then, having made a further quantity of plaster into a *thin* paste, pour it quickly on the fish and mould, taking care that the three holes be filled with the plaster. Should any air-bubbles occur during the pouring on of the plaster, they must be instantly dissipated with a soft brush or thin piece of wood. Having applied sufficient plaster to make a strong mould, let the whole rest until the moulds are quite hard, when they may be separated and the fish withdrawn. The upper mould will have three projections, corresponding to the holes in the lower mould, which will enable the operator to put the moulds together with facility and accuracy.

These moulds may now be placed in an oven until they are quite dry, and should then be put into a shallow vessel, containing melted wax, and allowed to remain therein until they are quite saturated; as soon as the moulds are cool they are ready to receive the plumbago, or other conducting medium.

Several holes should then be drilled in the edge of each mould, and a stout copper wire, bent at one end, be inserted in each hole, the terminations of these wires being well bound together, so as to prevent the mould from shifting from the wires. Several pieces of fine wire (jewellers' binding wire will answer this purpose very well) may be twisted round the conducting wire, and their ends be allowed to touch the surface of the mould in several places, in order to aid the deposit, which, when large surfaces are exposed, is apt to take place principally or entirely at the points of the conducting wires. Care must be taken that the portions of the mould to which the wires are attached be well coated with plumbago, and the edges of the mould should be scraped, in order to free them from any plumbago which may have been communicated to them; otherwise, when the deposit is obtained, it may be found difficult to separate the mould from the electrotype.

When the two halves of the fish are thus obtained in electrotype, the extraneous copper should be removed as before directed, and being filed until they will lay close together, the inner edges may be tinned with chloride of zinc and pewter solder, and being brought together, a blowpipe flame will soon complete the union. A perfect representation of the fish is here obtained, which may be either bronzed, plated, or gilt, by any of the processes hereafter to be described.

Moulds from any animal substances may be obtained by the above plan.

In some instances it may be advisable to make a mould of an animal in the elastic material before spoken of, in which case one half of the object may be imbedded in sand, and a cylinder of sheet tin made to surround the object, and being an inch or two higher it may be stuck in the sand. The elastic material is now to be passed into the cylinder, until it nearly reaches the top; it is allowed to remain until the composition has thoroughly set, when the metal rim may be removed, and the object separated from the mould. The other half of the object may be treated in the same way. The composition of wax and stearine may now be poured into each half of the mould, and plaster moulds may be taken from the wax models thus formed, which, being saturated with melted wax, may be plumbagoed and electrotyped.

Moulds from Vegetable Substances, may be generally taken in the same way as from animals. Leaves, sea-weeds, &c., may be thus copied:—Suppose we take a fern-leaf for example: let the back of the leaf be carefully imbedded in a paste of plaster of Paris, and with a piece of wood, guide the plaster so that it may fill up every crevice that is not to be copied. When the plaster is quite hard, melted wax may be poured over the leaf (which should be dusted over with plumbago previously, to prevent the wax from adhering), and allowed to remain until quite cold. The leaf and plaster should now be separated from the wax mould, which is then ready to receive the plumbago, &c.

Another good plan is, to brush over the back of the leaf with thin plaster, layer after layer, until it has received a good stout coating; this may now be imbedded in sand, and wax poured on as before.

Fern-leaves, sea-weeds, &c., may be imbedded in clay before the wax is applied to them.

The elastic moulding will also be found very useful in copying vegetable substances.

Gutta percha can seldom be applied with advantage to the copying of delicate objects of vegetable or animal nature, owing to the amount of pressure it requires to obtain an impression.

Having described the various moulding materials employed by electrotypists, we will proceed to the general applications of the art of electrotype.

Articles of glass may be coated with copper, by first covering them with a solution of gutta percha in turpentine or naphtha, or wax dissolved in turpentine; the article is then coated with plumbago, &c., in the usual way. The surface of the glass vessel may be rendered somewhat rough by submitting it to the fumes of hydrofluoric acid, but this is seldom requisite.

In some cases it will be found difficult to apply plumbago to a given surface, in which case the following mixture may be employed:—

Wax or tallow	1 pound.
Spirit of turpentine	1 pint.
India rubber	2 ounces.
Asphalte	1 pound.

Melt the wax or tallow, then dissolve the caoutchouc and asphalte in the turpentine, and add to the wax, stirring them well. Now pour in one pound of the following solution:—

Phosphorus	1 pound.
Bisulphuret of carbon	15 pounds.

Smaller quantities may be mixed up in the same proportions.

These substances being well blended together, objects to be electrotyped are brushed over with the composition, or, being attached to a wire, are dipped into it. A weak solution of nitrate of silver is next provided, containing about two pennyweights of silver to the quart of distilled water, into which the article is immersed until it assumes a black colour all over; it is then placed in clean cold water, and afterwards dipped in a solution of chloride of gold, washed again, and allowed to dry spontaneously. The object is now ready to be placed in the bath, where it will receive the deposit very readily.

The above method of rendering non-metallic substances conductors of electricity is particularly applicable to the coating of insects, flowers, and other delicate objects of nature.

Flowers, &c., may also be dipped in a rather weak solution of nitrate of silver, and then be exposed to the fumes of phosphorus under a glass; or the object *b*, after it has been dipped in nitrate of silver, may be placed in a bottle A charged with hydrogen, or phosphuretted hydrogen.

Daguerreotypes may also be copied by the electrotype process, thus:—A portion of the back of the daguerreotype is to be cleaned by scraping it, or by applying a single drop of nitric acid, which is then to be wiped off; a little chloride of zinc is now put on the clean spot, and a small piece of thin pewter solder. A thickish copper wire, having one end flattened, is now placed in the flame of a candle or lamp, and being brought in contact with the picture, the heat is to be continued until the solder runs. The back of the daguerreotype may now be coated with wax, and may then be placed in the bath to receive the deposit of copper. The electrotype will be found easily separable from the pictures, and it should be slightly gilt, in order to protect it from oxidation.

Another useful application of the art of electrotype is the invention of Mr. Palmer, termed by him *glyphography*, a description of which process we give herewith:—

"A piece of ordinary copper plate, such as is used for engraving, is stained *black* on one side, over which is spread a very thin layer of *white* opaque composition, resembling white wax, both in its nature and appearance; this done, the plate is ready for use.

"In order to draw properly on these plates, various sorts of points are used (according to the directions here given), which remove, wherever they are passed, a portion of the white composition, whereby the blackened surface of the plate is exposed, forming a striking contrast with the surrounding white ground, so that the artist sees his effect at once.

"The drawing, being thus completed, is put into the hands of one who inspects it very carefully and minutely, to see that no part of the work has been damaged, or filled in with dirt or dust; from thence it passes into a third person's hands, by whom it is brought in contact with a substance having a chemical attraction or affinity for the remaining portions of the composition thereon, whereby they are heightened *ad libitum*. Thus, by a careful manipulation, the *lights* of the drawing become thickened all over the plate equally, and the main difficulty is at once overcome: a little more, however, remains to be done. The depth of these non-printing parts of the block must be in some degree proportionate to their width; consequently, the larger breadths of *lights* require to be thickened on the plate to a much greater extent, in order to produce this depth. This part of the process is purely mechanical, and easily accomplished.

"It is indispensably necessary that the printing surfaces of the block prepared for the press should project in such relief from the block itself, as shall prevent the probability of the inking-roller touching the interstices of the same whilst passing over them; this is accomplished in wood engraving by cutting out these intervening parts, which form the lights of the print, to a sufficient depth; but in glyphography the depth of these parts is formed by the remaining portions of the white composition on the plate, analogous to the thickness or height of which must be the depth on the block, seeing that the latter is in fact (to simplify the matter) a *cast* or *reverse* of the former. But if this composition were spread on the plate as thickly as required for this purpose, it would be impossible for the artist to put either close, fine, or free work thereon; consequently the thinnest possible coating is put on the plate previously to the drawing being made, and the required thickness obtained ultimately as described.

"The plate thus prepared is again carefully inspected through a powerful lens, and closely scrutinised, to see that it is ready for the next stage of the process, which is to place it in a trough and submit it to the action of a galvanic battery, by means of which copper is deposited into the indentations thereof, and, continuing to fill them up, it gradually spreads itself all over the surface of the composition, until a sufficiently thick plate of copper is obtained, which, on being separated, will be found to be a perfect cast of the drawing which formed the *clichée*.

"Lastly, the metallic plate thus produced is soldered to another piece of metal to strengthen it, and then mounted on a piece of wood to bring it to the height of the printer's type. This completes the process, and the glyphographic block is now ready for the press.

"It should, however, have been stated previously, that if any parts of the block require to be *lowered*, it is done with the greatest facility in the process of mounting."

For the purpose of coating iron or zinc with copper, various solutions are employed.

1. Add to a solution of sulphate of copper a solution of cyanide of potassium, which will form a greenish precipitate; care must be taken to avoid breathing the fumes arising during this part of the process, as they are highly injurious. The precipitate is to be washed several times with cold water, and lastly dissolved with cyanide of potassium.

2. Pour into a solution of sulphate of copper, a solution of ferrocyanide of potassium, until no further precipitation takes place. Wash the precipitate as before. Cyanide of potassium will dissolve the precipitate. It is recommended to work this solution hot.

3. The solution which I have found to answer best for coating iron and zinc is composed of—

Carbonate of potassa		4 ounces.
Sulphate of copper		2 ,,
Liquid ammonia.	(about)	2 ,,
Cyanide of potassium		6 ,,
Water	(about)	1 gallon.

Dissolve the sulphate of copper in boiling distilled or rain water, and when cold add the carbonate of potassa and ammonia; the precipitate when formed is redissolved. Now add the cyanide of potassium, until all the blue colour disappears. A precipitate will be found at the bottom of the vessel, from which the clear solution may be separated by decantation.

The chloride or acetate of copper may be used instead of the sulphate, the former being preferable to the latter, but more expensive. Solutions thus made may be worked cold. Two cells of the battery described at page 7 will be found to answer admirably, for the purpose of depositing from these solutions.

Articles of iron which are to receive the deposit of copper should be previously soaked in a strong solution of caustic alkali, either soda or potassa, made by adding to either of these salts some recently slaked lime; the clear liquor proceeding from which is to be used for the purpose of removing any grease which may attach to the article, which is then to be well washed and immersed in a "pickle," consisting of—

Sulphuric acid	1 pound.
Hydrochloric acid	2 ounces.
Water	1½ gallon.

After the iron article has remained in this pickle for a short time, it may be removed, and well washed and scoured with sand and water, applied with a very hard brush.

Articles of zinc may be placed in the alkali, and then steeped in the following pickle:—

Sulphuric acid	1 pound.
Water	2 gallons.

After pickling, the articles may be scoured with sand if they require it, which is seldom the case, unless the work is old and greasy, in which case the brush and sand will readily remove any stains which may present themselves after pickling.

BRONZING.

When an electrotype is obtained, or a surface of iron or zinc coated with copper, a bronze appearance may be imparted by any of the following mixtures, which should be laid on with a soft brush, and allowed to dry; after which a somewhat harder brush should be briskly applied to the object, until it has become thoroughly brightened. Should the bronze, however, appear too uniform and want relief, a little of the composition should be rubbed off the raised surfaces, in order to give an effect of light and shade. This may be done to suit the taste of the operator.

As the bronzing mixtures are of different colours, and are to produce various effects, care should be taken never to apply any two of them with the same brush, without previously washing it.

Black Bronze.—Dissolve platinum in nitro-hydrochloric acid, and evaporate to dryness, or to crystallisation. Dissolve this in spirit of wine, ether, or water. A few drops of this solution may be mixed with any of the bronzing powders, such as crocus, sienna, rouge, &c. It is well to gently heat the article to be bronzed, previous to applying this composition. The projecting portions of the article may be lightened, if requisite, by applying a little liquid ammonia to them with a piece of chamois leather.

Brown Bronze.—Rouge, with a little chloride of platinum and water, will form a chocolate brown of considerable depth of tone, and is exceedingly applicable to brass surfaces, which are required to resemble a copper bronze.

Parisian Bronzes.

I.

Plumbago	1 ounce.
Sienna	2 ,,
Rouge	½ ,,

Add a few drops of hydrosulphate of ammonia and water.

II.

Chromate of lead	2 ounces.
Prussian blue	2 ,,
Plumbago	½ pound.
Sienna powder	¼ ,,
Lac carmine	¼ ,,

Add sufficient water to make a paste. To this may be added either chloride of platinum, or hydrosulphate of ammonia, according to the taste of the manipulator.

Another bronze may be made by mixing a little rouge, crocus, and hydrosulphate of ammonia in water; this should be applied several times, in order to give a body to the bronze.

Having given the principal facts connected with the electro-deposition of copper, sufficient I hope to enable the student to pursue the subject with ease and success, I now proceed to describe the various processes of Electro-Plating, in which I trust to present to the reader some useful practical information.

ELECTRO-DEPOSITION OF SILVER.

The most important of all the arts of electro-deposition is that denominated "electro-plating." This beautiful art is now practised to a vast extent in London, Sheffield, Birmingham, and Paris. Articles, chiefly made of German silver, are coated with fine silver, and thus, to a great extent, supersede the ordinary Sheffield and Birmingham plate; whilst old articles from which the silver has worn off can be replated, and thus rendered equal, and in some instances, superior to new.

Previous to the discovery of this art, when the silver had disappeared from the surfaces of plated articles by long usage, they became useless, as there was no process known by which the articles could be re-silvered.

Since the first introduction of the art, many have worked it with considerable success, and in the principal towns in England, Ireland, and Scotland, there are manufactories in which, annually, a vast amount of silver is deposited upon articles of various construction, and yet there is no superabundance of electro-platers; for I believe that if there were ten times the number, they would all do well, and for this reason: — the amount of plated goods now manufactured all over the kingdom far exceeds that made in the old days of Sheffield and Birmingham plate; and the silver which is deposited on these goods must be replaced as it wears off, in the progress of time, by the electro-plater. Again, many persons now use plated German silver goods in preference to silver, either owing to their superior beauty, their being less tempting to the marauder, or more economical to purchase. And when we bear in mind the vast quantity of electro-plate which is to be found in the hotels, restaurants, and private houses in the united kingdom—which is daily having its silver rubbed and scrubbed off, there is good reason to believe that the electro-plater's services will be extensively required, in proportion as the manufacture and consumption of electro-plate progresses.

There are many solutions employed in depositing silver upon various metals, from which we will select those most likely to succeed with the beginner and the practical man. The proportions of the materials used being the same in small or large operations, the manipulator may easily make up either of the following solutions in any quantity he pleases, from a pint to 1000 gallons or more.

Silver Solutions.—In making any of these solutions, perfectly *fine* silver must be employed; or, if it is desired to use standard or other impure silver, it will be better to purify the silver by first dissolving it in nitric acid; then add about one quart of cold water to the acid solution obtained from dissolving four ounces of silver. Now throw in a few pieces of sheet copper to precipitate the silver, and proceed as described at page 93. When the pure silver is thus obtained, it is to be again dissolved in two parts water and one part nitric acid.

Solution I.

Fine silver	. . . .	1 ounce.
Nitric acid	. . . . . . about	1 ,,
Water	. . . .	½ ,,

Put the silver carefully into a Florence flask, and then pour in the acid and water; place the flask on a sand bath for a few minutes, taking care not to apply too much heat, and as soon as chemical action becomes violent, remove the flask to a cooler place, and allow the action to go on until it nearly ceases; when, if there be silver still undissolved, the flask may be again placed on the sand-bath until the silver disappears. If, however, the acid employed has been weak, it may be necessary to add a little more. The red fumes formed when chemical action is going on disappear when the silver is dissolved or when the acid has done its work. If a little black powder be visible at the bottom of the flask, it may be taken care of separately, as it is gold. I have frequently found gold in the silver purchased of a refiner; in some instances more than sufficient to pay the expense of the acid employed.

The nitrate of silver formed during the above operation should be carefully poured into a porcelain or Wedgwood capsule, and heated until a pellicle appears on the surface, when it may be placed aside to crystallise. The uncrystallised liquor should then be poured from the crystals into another capsule, and heat applied until it has evaporated sufficiently to crystallise. When this is done, the crystals of nitrate of silver are to be placed in a large jar or other suitable vessel, and about three pints of cold distilled water added, the whole being well stirred with a glass rod until the crystals are dissolved.

A quantity of carbonate of potassa is now to be dissolved in distilled water, and some of the solution added to the nitrate of silver, until no further precipitation takes place. It is advisable occasionally to put a little of the clear solution in a glass, or test-tube, and to add a few drops of the solution of potassa, in order to ascertain whether all the silver is thrown down, or otherwise; as soon as the application of the alkaline solution produces no effect upon the solution of nitrate of silver, this operation is complete.

The supernatant liquor (that is, the fluid which remains above the precipitate) should next be carefully poured off the precipitated silver, and fresh water added; this is again allowed to settle, and the water poured off as before, which operation should be repeated several times in order to wash the precipitate thoroughly.

A quantity of cyanide of potassium is then to be dissolved in hot or cold water, and rather more than is sufficient to dissolve the precipitate added. In a few minutes the carbonate of silver will be dissolved by the cyanide, but in all probability there will be a trifling sediment at the bottom of the vessel, which may be separated from the solution by filtration, and preserved, as in all probability it will contain a little silver.

Sufficient water is now to be added to make one gallon of solution. Should the solution be found to work rather slowly at first, a little of the solution of cyanide may be added from time to time, as it is required: but it is preferable, in working a new solution, to have as small a proportion of cyanide as possible, otherwise the articles may *strip*, but more especially if they are composed of German silver.

When a silver solution has been worked for some length of time, it acquires organic matter, and is then capable of bearing, without injury, a larger proportion of cyanide.

It is necessary that the nitric acid employed for dissolving silver should be of good commercial quality, if not chemically pure, for if it contains hydrochloric acid (which is not an unfrequent adulteration), a portion of the silver dissolved will become precipitated in the form of a white flocculent powder (chloride of silver), and the success of the operation is thereby impaired.

Solution II.—To one ounce of silver, dissolved and crystallised as above directed, is to be added three pints of distilled water. The silver is to be precipitated from this by adding gradually a strong solution of cyanide of potassium. This must be done with caution, as an excess of cyanide will re-dissolve the precipitate. Should the operator, however, accidentally apply too much cyanide, a little nitrate of silver in solution may be added, the silver of which will be precipitated by the surplus cyanide. A portion of the solution should be placed in a wine-glass occasionally, and a drop of cyanide added, until no further effect is produced by this substance.

As soon as the precipitate (which is white) has subsided, the clear solution is to be poured off, and fresh water added, this being done several times, as before, to wash the precipitate.

Three pounds of ferrocyanide of potassium (yellow prussiate of potassa) may now be dissolved in water, and added to the precipitate. When the precipitate is dissolved, add sufficient water to make one gallon of solution, which should then be filtered before using. This solution is not very profitable to the electro-plater, as it requires fresh silver to be added frequently, owing to the fact that the anode, or silver plate, is not acted upon by the ferrocyanide, therefore the solution soon becomes deprived of its silver. It may be used, however, for experimental purposes.

Solution III.—One ounce of fine silver dissolved and treated as before, to which add three pints of distilled water. Precipitate the silver by adding a strong solution of common salt—an excess does no harm. A single drop of hydrochloric acid will show whether all the silver is thrown down or not. The white precipitate thus formed (which is chloride of silver) is to be washed as before.

A quantity of hyposulphite of soda is next dissolved

in hot distilled water, and a sufficient quantity added to dissolve the precipitate. Water is then to be added to make one gallon. This solution is decomposed by light, and should therefore be kept covered up, or in a dark place. It is not much used by electro-platers.

Solution IV.—One ounce of fine silver treated as before, and dissolved in three pints of distilled water. Precipitate with common salt, and wash, as above directed. Dissolve the precipitate with a strong solution of cyanide of potassium, taking care not to add much more than will dissolve the chloride of silver. Filter carefully, at least once through the same filtering paper and once through clean filtering paper, and then add enough distilled water to make one gallon of solution.

The above solution is very useful when it is desired to plate an article delicately white, but the silver is liable to strip when the burnisher is applied to it. This solution, however, may be employed with less fear of the work stripping, if it be used weaker, with a small surface of anode and feeble battery power.

Under all circumstances this solution is more applicable to surfaces which only require to be scratch-brushed, or which are to be left *dead*. Chased figures, clock-dials, cast metal work, &c., may be admirably plated with this solution.

Solution V.—One ounce of fine silver, as before, and the crystals dissolved in three pints of distilled water. Add strong solution of cyanide of potassium until no further precipitation takes place. If too much cyanide is added, it will re-dissolve the precipitate. Pour off the supernatant liquor and wash the silver as before. Now add strong solution of cyanide to dissolve the precipitate. Make one gallon with distilled water.

The solution should have a moderate excess of cyanide, and it must be filtered before using.

Solution VI.—A silver solution may be made by dissolving one ounce of silver as before. Dissolve the crystals in one pint of distilled water. Next be prepared with a large vessel full of lime-water, made by adding recently slaked lime to an ample quantity of water, which, it must be remembered, dissolves but a very small per-centage of lime. To the clear lime-water is to be added the solution of nitrate of silver, which will be converted into a dark brown precipitate (oxide of silver). When all the silver is thrown down, the clear liquor is to be poured off, and the precipitate washed as before. Now add strong cyanide of potassium solution to dissolve the oxide of silver, and make one gallon with distilled water.

This makes a very excellent solution, although it is somewhat troublesome to prepare.

Solution VII.—Dissolve in one gallon of water one ounce and a-quarter of cyanide of potassium, in a stone-ware or glass vessel. Fill a porous cell with some of this solution, and place it in the larger vessel; the solution should be the same height in both vessels. Then put a piece of sheet copper or iron, connected with the wire which proceeds from the zinc of the battery, into the porous cell. Place in the stone vessel a piece of stout sheet silver, which must be previously attached to the wire issuing from the copper of the battery. It is well to employ several cells alternated, for this purpose, when a large quantity of solution has to be prepared; that is to say, the zinc of one battery should be united by a wire with the copper of the next, and so on. In a few hours the solution in the larger vessel will have acquired sufficient silver, and the solution may be at once used. The porous cell is to be removed, and its contents may be thrown away.

In working this solution at first it is necessary to expose a rather large surface of anode, and small quantities of cyanide must be added occasionally until the solution is in brisk working order.

This is one of the best solutions, when carefully prepared, and is less liable to strip than many others.

Solutions of silver may be prepared by precipitating the silver from the solution of nitrate with ammonia, soda, magnesia, &c., &c., but for all practical purposes the solutions I., IV., V., VI., and VII., may, if carefully prepared, be depended upon.

When it is desired that the articles should come out of the bath having a *bright* appearance, a little bisulphuret of carbon is added to the solution. This is best done in the following manner:—Put an ounce of bisulphuret of carbon into a pint bottle containing a strong silver solution with cyanide in excess. The bottle should be repeatedly shaken, and the mixture is ready for use in a few days. A few drops of this solution may be poured into the plating bath occasionally, until the work appears sufficiently bright. The bisulphuret solution, however, must be added with care, for an excess is apt to spoil the solution. In plating surfaces which cannot easily be scratch-brushed, this brightening process is very serviceable. The operator, however, must never add too much at a time.

In making up any of the foregoing solutions the weights and measures employed are troy, or apothecaries' weight, and imperial measure, a table of which will be given at the end of this volume.

Having at command any of the solutions described, the operator may next arrange the battery. A plate *a, a, a,* or sheet of silver, is to be attached to the wire issuing from the copper of the battery *b,* and supported by a brass rod *d;* this may be done either by soldering them together or uniting them with a suitable binding screw; but the best plan of attaching the anode, or

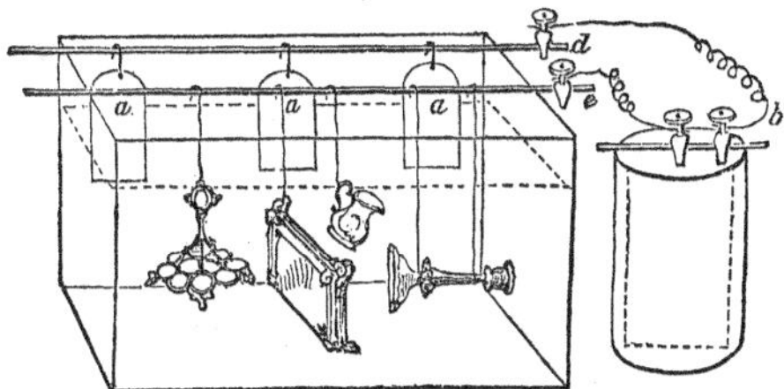

sheet of silver, to the copper wire is as follows:—Cut a strip to within half an inch or so; this strip may be united to the wire by a binding screw or soldered. If cast plates of silver are used, it is advantageous to have them cast with an extra piece, about three inches long at the corners, to attach the copper wire to.

The object in adopting either of the above arrangements is to prevent the copper wire entering the bath, as this is much impaired by allowing the copper to be immersed in the cyanide solution, whether deposition is taking place or not. Copper, if left in the bath for any length of time, even unconnected with a battery, will reduce a portion of the silver from the solution, an equivalent of the copper taking its place. This is especially the case when a large quantity of free cyanide is present.

A brass rod *e,* with a binding screw soldered or screwed on one end of it, is now to be attached to the negative wire of the battery. The articles to be coated may be suspended to this rod by pieces of clean copper wire; the wire used for this purpose may be rather thin, yet sufficiently strong to bear the weight of the articles. The thinner the wire is the less mark will be made upon the articles coated—a very important consideration in some cases, especially where spoons and forks are to be plated. This wire is termed "slinging wire." The size I generally prefer for spoons and forks is about $\frac{1}{32}$nd of an inch in thickness. The rods from which the anodes and goods to be plated are suspended must be kept quite clean and bright, by rubbing with emery cloth.

Preparation of New Work to be Plated.—German silver spoons and forks may be first placed in a hot solution of caustic soda or potassa (made by mixing recently-slaked lime with a concentrated hot solution of either soda or potassa, and allowing the lime to subside, the liquor is ready for use when further diluted), in order to remove any grease which may be upon them. A few minutes will effect this, as the caustic alkali very readily converts the small amount of grease generally on these articles into a soapy substance, easily removable by water. This process, however, is not indispensable; I seldom adopt it.

The spoons, &c., may now be well brushed with either powdered pumice-stone or powdered bath-brick (I prefer the latter) and water—a hard brush being applied to the purpose. This cleansing process is carried on until all the polish of the spoons is removed; and the fingers which hold the articles should be kept well charged with the powdered material, to prevent any grease or perspiration being imparted to the work. In cleaning spoons, it is advisable to begin at the inside of the bowl, and then to proceed to the other parts; lastly, going over the whole surface lightly, to render it uniform after the necessary handling it has been subjected to. A little practice will soon render the operator expert in these important details. The spoons, &c., are to be placed in clean cold water as soon as they are brushed, and are then ready for the bath. The slinging wires may now be attached.

When a solution is newly made, the work is apt to be irregularly coated at first, therefore it may be necessary to take the articles out of the bath about ten minutes after their first immersion, and to give them another slight rub with the brush and powdered material as before, when they should be again rinsed and placed in the solution.

If it is desired to give the spoons a very strong coating of silver, it is well, after a few hours' immersion, to remove them from the bath, and to submit them to the action of a lathe scratch-brush (consisting of a "chuck," with several bundles of fine brass wire attached to it, upon which beer or weak ale is allowed to run from a small barrel, with a tap to it, from above). This process will burnish down the white "burr," as it is called, and which consists of minute crystals of fine silver, and will prevent the coating from becoming *rough*. After the articles are scratched they should be rinsed in clean water, and again placed in the bath until done. The spoons may be lightly brushed over with moistened silver sand instead of being scratch-brushed, but the latter is preferable. When the goods have received the required coating they are again scratched, and can then be finished, either by the burnisher or polisher.

If the operator desires to know exactly how much silver is deposited on a given quantity of work, this may be done by weighing the article before and after immersion; or, by weighing the anode each time, he may form a tolerably correct estimate of the amount of silver deposited, for the anode generally supplies the solution with the amount of silver taken from it by the articles coated, that is to say, if all circumstances have been favourable.

When the articles are first placed in the bath, a sufficient surface of anode should be exposed (that is, immersed in the solution) to enable the goods to become whitish in the course of a few seconds. If they assume this appearance the very instant they enter the bath, the process is going on too quickly, and the articles will be liable to "strip." I regulate the speed of the operation of electro-deposition almost entirely by the anode, and I prefer exposing a small portion of this electrode at first, until the goods are uniformly covered, when the anode is lowered, little by little, until sufficient is exposed to carry on the operation with requisite speed. But the state of the solution and the battery must also be carefully attended to.

Large goods—for example, tea-pots, cruet-frames, tea-urns, &c., may be treated in the same way as spoons and forks, but care must be taken that no impression of the fingers be left on any of the plain surfaces, as in such case a roughness will occur at that part.

Preparation of Old Work to be Plated.—Sheffield or Birmingham plated cruet, soy, and liquor frames, &c., from which the silver has worn off, should first have the bottom separated from the wire, either by unsoldering or unscrewing, as the case may be. The bottom, if it is very rough, may be rendered smooth by means of emery cloth, or pumice-stone and water, and emery cloth afterwards. It may be finished with water-of-Ayr stone. The cruet-frame wire may generally be made smooth with emery cloth only.

As soon as the parts of the frame are smooth, the edges, feet, &c., may be brushed with a hard brush and powdered Bath brick, until all the interstices are quite clean. If there be any verdigris on any part of the frame it may be removed immediately by applying a few drops of hydrochloric acid ("spirit of salt") to the part. When the edges have been well brushed, the frame should be brushed all over in the same way, and it is then ready for the solution. But if the edges or mounts are lead ("silver edges" they are generally termed), it will be necessary to apply, with a rather soft brush, a solution made by dissolving four ounces of mercury in nitric acid, and adding about half-a-pint of cold water. This solution is to be lightly brushed over the lead mounts only; the article and brush are then to be well rinsed, and the brush and plain water again applied in the same way. The solution of mercury will turn the edges black, or dark grey, but the subsequent brushing will render them bright again. The frame is now to be well rinsed and is ready for the depositing bath. If, on its first immersion, any black spots exhibit themselves, the frame may be removed, again brushed over, and finally returned to the bath. If the edges do not receive the coating of silver as readily as the other parts, the solution may require a little more cyanide, or strengthen the power of the battery, or by increasing the surface of the anode this may be accomplished.

I have successfully coated these lead edges by apply ing a solution of sulphate of copper A, with a little free sulphuric acid in it, thus:—I dip one portion of the edge at a time in the solution of sulphate of copper, and with a piece of iron *b*, I touch the lead

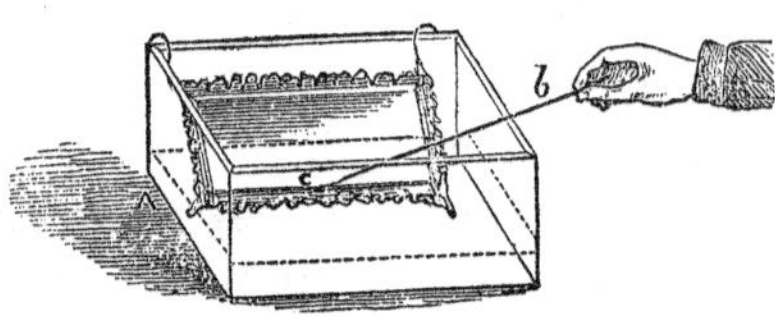

edge *c*, in solution, and this in an instant becomes coated with a bright deposit of copper. This is now rinsed, and the next part of the edge is treated in the same way, and so on. By this plan lead edges may be coated with great facility and certainty of success.

Generally, underneath the bottoms of cruet frames is a coating of tin; and as this metal is very troublesome to plate, unless in a solution made expressly for it, I prefer removing the tin, either by means of nitric or hydrochloric acid (the latter being rather a slow process), or with emery cloth and pumice; but nitric acid, employed with care, is the quickest plan.

When it is wished merely to whiten an article with silver, the amount deposited being of no consideration, solution No. 4, described at page 44, should be used. Let us suppose that a time-piece dial be the object to be whitened. The dial is first cleaned with a brush in the ordinary way, until all the old silver (if any) is removed; it is then rubbed with a piece of chamois leather and finely powdered Bath brick, slightly moistened; it is better to pass the leather over the surface *in circles*, so as to render the face as uniform as possible and to prevent the deposit from being *patchy*. The dial is then to be rinsed in quite clean water and suspended in the bath.

If the finger has been allowed to touch the face of the dial, it will be found that that part exhibits a dulness corresponding to the form of the skin of the finger, and it will be necessary to rub the dial as before with the chamois leather. The dial should be supported by the edges only. A few minutes' immersion will be sufficient to whiten a dial. When done, it is to be plunged into boiling water, and allowed to dry spontaneously, or be placed in perfectly clean box saw-dust.

Articles which are to be left with a dead-white surface, may be prepared in the same way, but they require to remain longer in the bath; in fact, till they assume the characteristic dead-white appearance. They are then to be placed in boiling water, and finally in box-dust, the latter being removed by means of a soft brush.

When it is necessary to whiten goods very quickly, the solution may be weakened with *hot* water, and the temperature raised to about 130° Fahr. The surface of anode exposed must be less than if the solution were to be worked cold. Moving the articles about in the solution occasionally, ensures uniformity and improves the whiteness by giving it a slight transparency.

When any of the solutions have been in use for some length of time, their conductibility may be augmented by adding a little cyanide of potassium. After the first few days the solution generally works better than when newly made; therefore it is not advisable to make any alteration in it until it begins to work rather tardily, when the additional cyanide may be added. I have invariably found that a solution that has been worked for several years has given better results than one recently made, and I have never yet been compelled to resupply the solution with silver; this is simply because I have taken care to work with a sufficient *surface* of anode, with battery power of feeble intensity, and with enough free cyanide in solution to cause the anode to yield as much silver to the bath as the plated articles have from time to time removed from it.

Iron is by no means an easy metal to coat with silver. It may, however, be successfully plated with care. The iron article should first be well cleaned and rendered free from rust, either by rubbing with emery cloth, or by dipping it into a pickle composed of

Sulphuric acid	2 ounces.
Hydrochloric acid	1 ounce.
Water	1 gallon.

It may remain in this pickle until the oxide or rust becomes easily removable by a brush and wet sand. If it be found, on removing the articles from the pickle, that the oxide does not brush off easily, it should be returned to the pickle-bath. When the surface is merely rusty, strong hydrochloric acid alone will remove the rust and render the article at once clean and ready for the sand-brush. The articles when cleaned and well rinsed, may be placed in the alkaline solution of copper bath, described at page 35, and allowed to remain until they have received a slight coating; they may then be rinsed and placed in the silvering bath; or the articles may be electro-brassed by any of the processes hereafter to be described, and then immersed in the plating bath.

It is better to deposit a coating either of brass or copper upon an iron surface, to insure success. Copper will adhere well to iron, but silver will not, therefore copper acts the useful part of a "go-between," thereby preventing the disagreement that might arise were two metals, so antagonistic to each other as silver and iron, allowed to come in contact.

The solution in which iron is to be plated should be weakened with about fifty per cent. of water.

Britannia metal, pewter, and all combinations of lead and tin, are best plated in a solution containing a good deal of free cyanide. Deposition should be suffered to take place quickly at first, so as to insure the deposit going well all over the article. A larger surface of anode, also, must be exposed than would be required for German silver work—probably three times the surface.

The battery power must be energetic, but not too intense. Two 4-gallon cells of the battery described at page 10 will be sufficient for objects of considerable size. Articles made of Britannia metal, &c., should not be disturbed while in solution. They may, however, be shifted now and then so as to expose a fresh surface to the anode, for the sake of causing uniformity of deposit, but it is not advisable to let the solution be agitated more than is absolutely necessary. This caution, however, is chiefly applicable to the period when the articles are first immersed in the bath.

The goods may be prepared for plating by brushing them over with silver sand and water, with a moderately hard brush, instead of the powdered Bath brick used for other metals. The articles may be cleansed from grease by placing them for a few minutes in a hot solution of caustic soda.

If the articles, when they have been placed in the plating bath for a few moments, present an unequal surface, it is advisable to remove them, and have them brushed over again as before; then, after well rinsing, they should be quickly returned to the bath and allowed to remain without further disturbance if possible.

The readiest mode of plating articles of lead, tin, or zinc, is to previously cover them with a film of brass, in the brassing bath: or with copper, in the alkaline coppering bath. Either brass or copper will adhere firmly to the surface of these metals, and the silver may be deposited on the coppered or brassed surface with perfect facility. After the articles have been removed from the copper or brass bath, they should be well scratch-brushed and rinsed before placing in the plating bath.

When an article has been plated and is found to strip or blister in many places, it will be necessary to remove all the silver from the surface and plate it again more carefully. There are two kinds of *blistering;* the first is the non-adherence of the silver to the article coated; the second, the blistering or *doubling* of the metal of which the article is composed. When the blistering is of the first kind, the article, after the silver has been removed by the process described below, may be rendered smooth with water-of-Ayr stone previous to being replated; but if the metal itself has been blistered when the burnisher was applied to it, the blisters must be scraped, buffed, or filed down, and the surface made smooth in the ordinary way.

The silver is removed (or "stripped") from articles thus:—Put some strong sulphuric acid into a stone jar or enameled saucepan, and add a few crystals of nitrate of potassa (saltpetre). Apply heat until the nitrate is dissolved. When this solution is very hot the articles are to be immersed, and occasionally moved about until the silver becomes entirely dissolved from the surface. The removal of the silver will become manifest by the metal beneath being exposed at the edges. The operation must then be closely watched in order to prevent the articles remaining longer in the solution than is absolutely necessary. If the above solution does not remove the silver quickly, more nitrate of potassa must be added from time to time, and the heat augmented if required.

When a quantity of things have been stripped in the solution it will begin to work slowly, and a mass of crystals will be found at the bottom of the vessel as it cools. It will be better now to add a quantity of cold water to the solution, and to immerse in it some pieces of zinc, which will throw down the silver in a metallic state, but in minute crystals of a greyish colour. By dropping a little hydrochloric acid into the solution the operator will be able to judge whether all the silver is deposited or not. The acid will form a white precipitate so long as there is any silver in solution. When all the silver is precipitated, the supernatant liquor may be poured off carefully and fresh water added to wash the precipitate, which process should be repeated several times in order to render the silver as clean as possible. The pieces of zinc should be removed before the final washing of the silver. The silver may now be dried and put into a crucible (being previously mixed with a little *dry* powdered potash), and heated in a furnace until all the metal is gathered into a button. During the process of fusion, a few crystals of nitrate of potash may be carefully dropped into the crucible. The silver thus obtained will be perfectly fine.

Or the silver may be precipitated from the *stripping* solution by means of common salt, when chloride of silver will be formed, which may be dried and fused in the same way as above, previously being mixed with a little soda or potassa, to form a flux.

The articles may be stripped in the plating bath, by suspending them in the place of the anode, but this plan is apt to injure the solution, by imparting to it a portion of the copper or other metal of which the article was made. Or the silver may thus be precipitated upon the end of a wire, enclosed in a linen bag to collect the small granules of silver which will fall from the end of the wire. A powerful current would strip the silver off in a very short time. A small quantity of solution containing excess of cyanide should be kept for this purpose alone.

Deposition of Silver upon Non-Metallic Substances.—Silver may be deposited in the same way as copper by the electrotype process, but as cyanide ot potassium rapidly dissolves wax, it will not be advisable to employ moulds made of that material. Gutta percha is better, but even this substance is acted upon by cyanide of potassium. Moulds made of fusible metal are, however, more suitable for this purpose. When the cyanide solution of silver is employed for depositing upon or for copying non-metallic substances, the bath should have at least six times as much silver as that required for the ordinary process of plating.

A solution of nitrate of silver, if not too strong, will answer well for depositing upon plumbagoed surfaces, but this solution must not be employed for depositing upon copper or other metallic surfaces.

The backs of copper moulds which are to be coated

with silver should be covered with some material to prevent the silver being deposited on those surfaces; for this purpose it has been recommended to boil a little pitch in a strong solution of potassa, which will form a sediment. Some of this sediment is added to a quantity of melted pitch, at which time a violent action ensues, white fumes being evolved. Allow this action to subside, and the resulting material will be ready for use. Melted gutta percha will also answer tolerably well for protecting the backs of metallic moulds.

There are many other practical points in the electro-deposition of silver, which we deem it advisable, for the reader's convenience, to give in the form of an Appendix at the end of this work, to which he will prudently refer when desirous of practising the art of electro-plating, &c.

ELECTRO-DEPOSITION OF GOLD.

In importance, electro-gilding is second only to the art of electro-plating; and it is carried on in much the same way. The solutions of gold, however, should be generally worked *hot;* hence the operation of gilding is conducted in a much shorter space of time than is required for plating. An article may be well and strongly coated in a few minutes, whilst it would require several hours to electro-plate an article well.

There are many forms of solution in use amongst electro-metallurgists, all of them varying in the proportion of gold to the gallon of water, and in the amount of cyanide employed. These solutions are all of them easily made, and any of them can be well worked by a skilful operator. Some gilders use five or six pennyweights of gold to the quart of solution—others as much as eight or ten dwts.; but I have generally found that a solution containing less gold will give better results than one richer in the metal, independent of the advantage in point of economy. I have observed that a bath containing five or six dwts. of gold to the quart of water, and the necessary proportion of cyanide, and worked with several united cells of Smee's battery, has required a much larger surface of anode to be exposed to a given surface of negative electrode (that is, the article to be gilt) than would be required to gild an article in a solution containing one and a half dwt. to the quart of solution worked with a single cell of a constant battery. Hence I infer that the weaker solution is the better conductor of the two.

Gold Solutions.—Solution I.—Dissolve in a Florence flask one pennyweight and a half of fine gold in two parts hydrochloric acid and one part nitric acid (*aqua regia*), applying gentle heat to accelerate chemical action. When the gold is all dissolved, pour the chloride of gold thus formed into a porcelain capsule and apply moderate heat until all the acid is evaporated. A red mass will result. It is advisable, when the acid is nearly expelled, to move the capsule round and round, so that the liquid may be dispersed over a large surface of the vessel. It will be found that the liquid will cease to flow when the acid is expelled, at which period the operation is complete. If too much heat is applied the gold will become reduced to the metallic state, which may be known by the red mass acquiring first a yellow tinge, and next a gold bronze will be observed at the bottom of the capsule. In such a case it will be necessary to add a little more of the mixed acids in the same proportion as before, which will at once redissolve the reduced gold.

When the acid has been driven off the chloride of gold, about half a pint of cold distilled water is to be added, which will at once dissolve the chloride, forming a bright straw-coloured solution. Allow this to subside for a few minutes, as in all probability there will be a small amount of white precipitate at the bottom of the vessel, which is chloride of silver; the solution of gold must be carefully poured off from this precipitate, as it is soluble in cyanide of potassium, and its presence in the resulting solution may be prejudicial. A little distilled water may be poured into the capsule, to rinse away all the gold, taking care not to allow the sediment to come away with it, when transferring the rinsings to the solution of gold.

A little strong solution of cyanide is now added, gradually, to the solution of gold, and the whole stirred with a glass rod. The gold solution will instantly lose its yellow colour. A brown precipitate is formed by the solution of cyanide, and this solution must be added drop by drop until it produces no further effect upon the clear solution. The supernatant liquor is now to be carefully poured off, and fresh water added several times to wash the precipitate of gold—taking care not to waste any of the precipitate nor to add more cyanide than is *absolutely* necessary. When the precipitate is sufficiently washed, more of the solution of cyanide is added, which will at once dissolve the precipitate, forming a clear solution. The cyanide should be added in excess, say about twice as much as may be required to dissolve the precipitate. The concentrated solution of cyanide of gold thus obtained is placed over the fire or on a sand-bath until it is evaporated to dryness, when it may be again dissolved in cold water and filtered for use. Lastly, enough boiling distilled water is added to make one quart of solution, and a little additional cyanide added if the solution is found to work too slowly at first—but it is better not to use more cyanide than is necessary, otherwise the anode will become rapidly consumed and the gilding be of a "foxy" colour.

Solution II. Dissolve one and a half dwt. fine gold as before, and evaporate to dryness. Re-dissolve in half a pint of distilled water and precipitate the gold with ammonia, taking care not to add more ammonia than is necessary. Pour off the supernatant liquor and wash the precipitate as before. Now add sufficient cyanide of potassium to dissolve the precipitate. Evaporate to dryness, and re-dissolve with cold distilled water. The solution is then to be filtered, and distilled water added to make one quart. A little cyanide is to be added occasionally, as required.

Solution III. Dissolve one dwt. and a half as before, and when the half pint of solution of chloride is obtained, precipitate the gold with hydrosulphate of ammonia. A copious black precipitate is formed, which must be allowed to subside, and this substance then washed as before directed. Dissolve the precipitate with a lump of cyanide—say about half an ounce, or rather less; and evaporate to dryness. Then add water to make one quart.

Solution IV. Dissolve the same quantity of gold as before, but without evaporating the acid. Add a quantity of calcined magnesia, which will precipitate the gold in the form of an oxide. To the oxide add sufficient concentrated nitric acid (applying heat at the same time) to dissolve the magnesia, when the oxide will be left in the form of a precipitate, which is to be well washed, and then solution of cyanide added to dissolve it as before. Evaporate and make one quart of solution with distilled water.

Solution V. Dissolve one ounce of cyanide of potassium in one quart of nearly boiling distilled water. About half fill a "porous cell" with the solution, and stand it in the vessel containing the bulk of the solution. Attach a piece of sheet copper to the wire issuing from the zinc of the battery, and place it in the porous cell. Put a piece of sheet gold, attached to the copper of the battery by a wire, in the outer solution, and allow the whole to remain in action until the solution has acquired about one pennyweight and a half of gold, which may be ascertained by weighing the gold before and after immersion. The porous cell may now be removed and its contents thrown away. The solution is now ready for use.

These solutions should be worked at a temperature of about 130° F., with one cell of a constant battery.

The solution of gold may be heated either in an enameled saucepan, or in a glass vessel placed in an iron pan containing water. The operator now proceeds to arrange his battery. The wire which issues from the copper of the battery is to be attached to a piece of fine gold, which may conveniently be done by soldering. The article to be gilt is to be suspended to the wire proceeding from the zinc of the battery.

Preparation of Articles to be Gilt.—Silver goods, such as cream ewers, sugar bowls, mugs, &c., should be well scoured inside with hot soap and water and silver sand, and if they are at all greasy, a little caustic soda may be applied to them first. Or the mugs, &c., may be well scratch-brushed and then rinsed with boiling water. The insides only of these vessels are generally required to be gilt, in which case the outsides may be wiped dry before gilding. The negative wire (from the zinc of the battery) is to be attached to the handle of the vessel. The plate of gold is now to be carefully suspended in the centre of the mug, taking care that it does not touch the vessel; and the gold solution may be poured into the mug by means of a jug or other suitable vessel, until it reaches the upper edge. If it is desired to gild the extreme edge, the solution may be guided over it with a piece of wood or glass rod. In about five or six minutes the vessel will be sufficiently gilt, when the anode may be removed, the negative wire detached, and the solution poured into the bath. The article is at once to be rinsed with hot water, and may be scratch-brushed and burnished in the ordinary way. When cream ewers, &c., are so constructed that the solution will not reach the lip, &c., without overflowing, it is advisable to slightly tilt the vessel so as to cover as much of it as possible, and when it is gilt the lip may be dipped into a little gold solution, being attached to the battery the while; but in this case the outside of the lip will also receive a deposit. This may be prevented by coating the outer surface of the vessel with the composition which we have already described, p. 59. Vessels which are to be gilt inside only, should be placed on a plate or dish to collect any solution which may run over.

Silver brooches, pins, rings, thimbles, egg, salt and mustard spoons, &c., merely require to be scratch-brushed before gilding. After they have received the required deposit, they are again brushed, and if the colour be a little too pale or too red, the articles should be immersed in the bath again *for an instant*, and then plunged into boiling water, when they will assume a beautiful fine gold colour. When well rinsed in hot water, the articles are to be placed in box saw-dust, which may sometimes be advantageously kept hot for this purpose, in order to dry the goods as speedily as possible; but care must be taken that the box-dust be not allowed to char or burn, otherwise it will stain the articles.

Goods which are made of copper or brass entirely, may be dipped into nitrous acid ("fuming nitric acid" or "dipping acid") for a moment, and instantly plunged into clean cold water; after which process they should be again rinsed in fresh water, and at once placed in the gilding bath. Or such articles may be merely scratch-brushed, rinsed, and then placed in the bath.

If, when first put in the bath, copper or brass goods receive the deposit too quickly, the anode should be raised a little out of the solution, so as to expose a smaller surface, and the articles should be moved about a little, by which uniformity of deposit will be secured. In fact, it is advisable always to give the articles a gentle motion when first placed in the bath, until they have received a slight coating, when they may be allowed to remain steady until finished; but when it is required to deposit a stout coating, it will be advantageous to move them occasionally, to prevent the deposit taking place unevenly.

When goods are made of either copper or brass, with mountings of another metal, or if they have been previously plated or gilt, greater care must be observed, otherwise some parts will receive the deposit favourably while others will scarcely be coated at all. This applies more especially to goods which have mountings pewter-soldered upon them, which is frequently the case in common jewellery. In this case all the surfaces will receive the deposit but the solder, which, being a bad conductor of electricity, and more electro-negative than the other metal to which it is attached, will receive the deposit but tardily, if at all. I have frequently found that the smallest speck of pewter solder which has happened to be upon a brooch which I had to gild, has compelled me to deposit at least three times as much gold as the article required before I could cover the speck of solder; and in many instances not even then would deposit take place upon the offending spot. Having tried to amalgamate the solder with the gilt surface by means of nitrate of mercury, nitrate of silver, and both combined and alternately applied; and having scratch-brushed the tardy spot until I was heartily sick of pewter solder and everything which it contaminated, I at last hit upon a plan by means of which I have ever since been enabled to gild pewter solder with ease and certainty. I placed a single drop of an acid solution of sulphate of copper upon the solder spot, and then touched it with a piece of steel: in an instant the solder and surrounding surface received a bright deposit of copper (which could be strengthened by repeating the operation several times). The moment the article was placed in the gilding bath the spot became coated; in fact—copper being easier to gild than gold—this spot received the deposit in preference, so that my difficulty was speedily and satisfactorily overcome. Many electro-gilders, I have no doubt, will find the above plan relieve them from a considerable amount of annoyance. Generally speaking, however, when the operator finds a difficulty in gilding pewter solder, it is owing to the bath requiring cyanide, or the exposure of a larger surface of anode; or may be the battery power is weak.

Instead of the above plan of coating pewter solder, the manipulator may put a drop of concentrated solution of silver upon the solder, as before, and, on touching the part with a piece of fine wire the solder will be coated with silver in an instant. I prefer the former plan, however, since copper receives the deposit of gold more readily than silver.

In gilding cheap jewellery, French and Birmingham fancy goods, and articles which are not required to have more than a *coloured surface* given to them, I have found it an economical plan to gild with a copper anode, and as the gold becomes exhausted from the solution, to add more gold from time to time, thus working from the solution instead of from the anode. By this arrangement, the operator is sure not to deposit more gold upon his work than is consistent with the scale of remuneration for doing the same.

Generally, it is only necessary to scratch-brush this class of goods; then having rinsed them in boiling water, they are to be dipped into the solution for an instant; a few seconds only being required to give the goods a beautiful colour.

Silver filigree brooches, &c., must be well scratched, dipped in the bath for a moment and then rinsed and scratched again: on immersion in the bath the second time they will become more uniformly gilt. The surface of anode must be adequate and the battery power brisk, or the filigree work will not receive the deposit uniformly. The solder of filigree work is generally the most troublesome to gild, but if the current and other circumstances are favourable, the deposit should take place all over at once; a want of cyanide in the solution is a principal cause of difficulty in gilding filigree work. Gently moving the articles about in the solution, greatly adds to the beauty and uniformity of the result, if all other matters are favourable. The solution for gilding filigree goods should contain rather more gold than that for ordinary work, and the surface of anode exposed must be greater.

Metal pins, brooches, rings, &c., should be either "dipped" as before recommended, or well scratch-brushed before gilding. This class of goods may be done on a large scale, in a porcelain vessel like a colander, suspended in the bath. The goods, being placed in this, merely require to be touched, and occasionally stirred about by the negative pole of the battery, so as to cover those parts which have been in contact with each other.

Army accoutrement work, sword-mountings, &c., should first be prepared by cleaning with silver sand, soap and water, applied with a hard brush. The article may then be scratch-brushed, and placed in the bath until it has *nearly* acquired a sufficient coating, when it is to be removed, rinsed in warm water, and those parts which are required to be left *dead*, gently brushed over with powdered pumice or Bath-brick. The article is then returned to the bath and allowed to remain until finished. As soon as it has become sufficiently gilt, the plain surfaces may be scratch-brushed, and then burnished.

I have induced my workpeople to burnish gilt-work with four-penny ale, instead of soap and water, and it has been considered by them a great improvement, since the burnisher seems to glide over the surface of the work with greater ease and smoothness, more especially when the gilding has been what is termed "hard" or "scratchy." Vinegar has sometimes been applied to this purpose, but not, I think, with such success.

In gilding German silver, the solution may be worked at rather a lower temperature, the solution weakened, and a less surface of anode exposed. German silver has the power of reducing gold from its solution in cyanide (especially if the solution be strong) without the aid of the battery; as also will brass receive a coating of silver in the plating-bath without the use of the current, therefore the solution should be weaker—in fact, so weak that the German silver will not deposit the gold *per se*; otherwise the deposit will take place so rapidly that the gold will peel off when being burnished, or even scratch-brushed.

When iron or steel goods are to be gilt, they should be first rendered free from grease, by being immersed in a solution of caustic soda or potassa; they are then to be well scratch-brushed—in fact, until they have acquired a slight coating of brass, from the wires of the brush. If sour beer is used for this purpose it will greatly facilitate the operation. The article should then be placed in the bath for an instant, then well scratch-brushed and dipped again. The solution employed for iron or steel should be much weaker than for any other metal. I would recommend the following:—

Ordinary solution	4 fluid ounces.
Water	20 ,,
Cyanide of potassium, about . . .	2 drachms.

This solution may be worked rather warm, but not so hot as the ordinary solution. Weak battery power should be employed, and small surface of anode, and deposition must be allowed to take place very slowly at first.

By scratch-brushing iron or steel articles with vinegar or dilute hydrochloric acid, a very good and adhesive coating of copper may be obtained upon the surface of the article, but the employment of the latter must be done with caution or the operator's clothes may be injured; a few drops of acid, however, to the pint of water is all that will be required.

The best method of preparing steel or iron articles for gilding, is to coat them with copper or brass in the same way as that recommended for plating these metals. Many steel articles, which only require a trifling deposit of gold, may be gilt without any further preparation than merely rinsing them in hot water. The articles then receive a momentary dip in the bath, and, being sufficiently gilt, are rinsed in hot water and dried quickly in hot box-dust, or in an oven.

Steel surgical instruments must be gilt with great care, in order that the edges be not rendered blunt by the operation. These articles should be placed in the bath without any preparation, as coating them with copper or brass, and then gold, may involve too much handling. A slight deposit is all that is necessary to protect the steel instrument from rust or corrosion.

Steel or iron keys should be first well scratch-brushed, dipped into the bath for a moment, and then brushed again; lastly, allowing them to remain in the bath until sufficiently coated. These may be finished either by burnishing or polishing.

ELECTRO-DEPOSITION OF BRASS AND BRONZE.

It is far more difficult to deposit an alloy of two or more metals than one only; and this difficulty becomes greater when we require to deposit, as an uniform alloy, two metals whose electrical conditions are of an opposite character, as zinc and copper. From a solution consisting of zinc and copper in the proportions to form ordinary brass, it is easy to deposit the zinc alone, or the copper alone, by increasing or diminishing the power of the current, or by raising or lowering the anode; that is to say, by increasing or diminishing the surface of anode exposed to a given surface of object to be coated.

The difficulty in regulating all circumstances, so that an uniform result might be obtained by the operator, and so that the process of electro-brassing might be depended upon, has, in many instances, caused this useful art to be abandoned altogether by the manufacturer.

Many processes of electro-brassing have been published and patented in this country and on the continent, but all of them have the disadvantage of being more or less troublesome and uncertain to manage, even though the operator be a person well skilled in electro-deposition. But I think that several of these processes may be rendered commercially valuable if the solutions in the first instance are mixed by persons acquainted with chemical laws. Again, there would be less liability to failure, if the power of the current employed was always regulated by the surface of goods to be coated; the amount of anode, also, being regulated by the same. If, on the contrary the battery-power be too weak, or in excess, either the copper on the one hand, or the zinc on the other, will be deposited alone.

In giving the various processes of electro-brassing, I may inform the reader that several of them are patented, and consequently cannot be employed for commercial purposes without the permission of the respective patentees.

I. De Salzede's Patent Processes.

I.

Cyanide of potassium	12 parts.
Carbonate of potassa	610 ,,
Sulphate of zinc	48 ,,
Chloride of copper	25 ,,
Nitrate of ammonia	305 ,,
Water	5000 ,,

Dissolve the cyanide of potassium in 120 parts of the quantity of water above specified, and then dissolve the carbonate of potassa, sulphate of zinc, and chloride of copper in the remaining water, raising the temperature to about 150° F.; and as soon as the salts are well dissolved add the nitrate of ammonia, frequently stirring until the latter is dissolved. The solution may now be allowed to stand for a few days, in order that the sediment formed may become precipitated, when the clear liquor is to be drawn off, and is ready for use.

II.

Cyanide of potassium	50 parts.
Carbonate of potassa	500 ,,
Sulphate of zinc	35 ,,
Chloride of copper	15 ,,
Water	5000 ,,

This solution may be made up in the same way as No. 1.

3. Bronzing solution.

This solution is the same as No. 1, excepting that 25 parts of chloride of tin are used instead of the sulphate of zinc.

4. Bronzing solution.

In this solution 12 parts of chloride of tin are employed instead of sulphate of zinc in the second brassing solution. This latter solution Salzede works at a temperature not exceeding 97° F.

The above solutions are to be worked with a brass anode, and with an active battery of two or more cells —Bunsen's battery being preferable to any other. The current of electricity employed in electro-brassing must have a brisk intensity—the quantity also being considerable.

The above solutions work very well at first, but they soon get out of order, owing to the irregular action of the cyanide upon the brass anodes, which readily attacks the copper, whilst the zinc frequently remains upon the surface of the anode in the form of a white paste. Hence the character of the solution soon becomes altered.

II. Brass Solution.

Acetate of copper	5 ounces.
Potassa	4½ pounds.
Sulphate of zinc	10 ounces.
Liquid ammonia	1 quart.
Cyanide of potassium	8 ounces.

Dissolve the acetate of copper, which should be previously pulverised, in half a gallon of water. Add 1 pint of the liquid ammonia, and then dissolve the sulphate of zinc in 1 gallon of water, the temperature of which should be raised to about 180° F. When the zinc is dissolved, add the remaining pint of liquid ammonia to the solution, which should be well stirred immediately, in order to insure its perfect mixture with the sulphate of zinc.

Dissolve the potash in one gallon of water. Lastly, dissolve the cyanide of potassium in one gallon of hot water, and then mix the ingredients in the following order:—The solution of copper to be added to that of zinc; now add the solution of potash and cyanide. Stir the whole well together, and allow the mixture to digest for an hour or so, stirring occasionally. Add water to make altogether 8 gallons of solution.

The above solution must be worked with active battery-power and a brass anode—milled brass being preferable. The anode should be well cleaned before immersion. A little liquid ammonia may be added from time to time, and also a small portion of cyanide when the solution works slowly. The anode must be kept clean. I have also found it advantageous to add a little arsenious acid to the solution, which improved the character of the deposit, by rendering it brighter and less crystalline. The arsenious acid, however, does not at first appear to make much difference, but after a while the improvement becomes manifest. I generally apply the arsenic by mixing it with a strong solution of cyanide of potassium. About one ounce to the above solution will be sufficient at first, and the quantity may be increased by degrees.

III.

Acetate of copper	10 pounds.
,, zinc	1 pound.
,, potassa	10 pounds.

Dissolve the above substances in 5 gallons of hot water, and add cyanide until a precipitate is formed, which, upon adding more cyanide, becomes again dissolved. An excess of cyanide must be added. The patentees of this process (Messrs. Russell & Woolrich) use either a brass anode, or one of brass and another of copper at the same time.

IV. Bronze Solution of M. Brunel & Co.

Chloride of copper	1 pound.
Carbonate of potassa	25 pounds.
Sulphate of zinc	2 ,,
Nitrate of ammonia	12½ ,,

The chloride is to be dissolved in half a gallon of water; the carbonate of potassa in 6 gallons of water; the sulphate of zinc is to be dissolved in half a gallon of hot water. These three solutions are to be mixed together. Now add the nitrate of ammonia, and blend them all together by stirring well for a few minutes. Make about twenty gallons, by adding cold water.

This solution is to be worked in the same way as either of the above.

The above solution much resembles M. Salzede's process, and is prone to get out of order owing to the fact that the anode does not supply the solution with metal as fast as it is deprived of it by the articles coated. Unless the solvent employed will readily attack and dissolve the zinc of the anode, the solution must soon lose its proportion of this metal. The liquid ammonia used in one of the above processes seems to effect this more satisfactorily than an excess of cyanide. I have invariably found that in any of the above processes, the employment of a liberal amount of liquid ammonia has kept the anodes clean, and enabled the solution to give better results in every respect. The white salt of zinc formed upon the surface of the anode is soluble in this menstruum, but sparingly so in cyanide of potassium. The ammonia and cyanide being in the solution in abundance, will keep the anode clean, without which the action soon ceases altogether.

V. Newton's Process consists in forming solutions for depositing alloys of copper, tin and zinc, and also, for depositing brass and bronze.

The patentee mixes chloride of zinc with chloride of ammonium, sodium, or potassium dissolved in water.

Acetate of zinc in solution mixed with acetate of ammonia, potassa, or soda.

In making up a brassing solution, Newton adds to either of the above solutions a proportion of a corresponding salt of copper—for instance, with the acetate of zinc he would unite the acetate of copper, and so on. He employs various other salts of zinc, with the corresponding copper salt, for the same purpose.

In making a bronzing solution Mr. Newton dissolves the double tartrate of copper and potassa, and double tartrate of the protoxide of tin and potassa, with or without the addition of caustic potassa. He deposits an alloy of zinc, tin and copper, by using a solution composed of the following substances:—double cyanide of copper and potassium; zincate of potassa and stannate of potassa; the zincate of potassa he forms by fusing oxide of zinc with caustic potassa, and the stannate of potassa either by fusing oxide of tin with caustic potassa, or by dissolving it in a solution of potassa.

For an electro-brassing solution the patentee employs a solution composed of a given quantity of oxide of copper, dissolved in an excess of cyanide of potassium;

oxide of zinc and a little liquid ammonia are then added, and the solution heated to 120° Fah., to 140° Fah. Water is then added, in sufficient quantity to allow the solution to contain about 3 oz. of the oxides to the gallon—*i. e.*, 2 of zinc to 1 of copper to form brass.

VI. Brassing Solutions.

I.

Cyanide of potassium	1 pound.
Carbonate of ammonia	1 ,,
Cyanide of copper	2 ounces.
,, zinc	1 ounce.

Dissolve in one gallon of water. The temperature to be raised to 150° Fah.

II.

Cyanide of potassium	1 pound.
Carbonate of ammonia	1 ,,

Dissolve in one gallon of water. Attach a large brass anode to the positive wire of a battery, and apply a small surface of cathode or negative electrode—say a strip of brass. The temperature should also be 150° Fah. By this arrangement the anode dissolves, supplying the solution with metal. The exact quantity which the solution has taken up may be ascertained by weighing the anode before and after immersion.

VII. Brunel gives another formula for a brassing solution :—

Carbonate of potassa	10 pounds.
Cyanide of potassium	1½ pound.
Sulphate of zinc.	1¼ ,,
Chloride of copper	10 ounces.
Water	12½ gallons.

The best way of making up the above solution is to dissolve all the ingredients in separate vessels ; then to add to the sulphate of zinc and chloride of copper a portion of the solution of carbonate of potassa. Now add sufficient liquid ammonia to dissolve the respective precipitates at first formed, when the solution of cyanide and the remainder of the carbonate of potassa may be poured in, and water added to make altogether 12½ gallons. This solution must be worked with a large brass anode, and a brisk battery of two or more Bunsen's cells. The solution should stand for some hours before using it, when it may be separated from any sediment which may remain at the bottom of the vessel in which it is made.

The above solution will require to be replenished from time to time with a little cyanide of potassium and liquid ammonia, in order to keep the anode free from the white salt of zinc, which would otherwise form upon its surface. Arsenious acid improves this solution; and I have found that a little chloride of tin, dissolved in caustic potassa, tends to toughen the deposit.

Iron, lead, zinc, tin, and alloys of lead, &c., will not all receive an equally good coating of brass if placed in the bath at the same time. No two metals of different characters should be immersed together; and, indeed, different solutions should be employed for each metal or alloy.

Cast iron requires a solution containing a greater per-centage of metal than zinc or its alloys; whilst zinc will receive a good deposit when but little metal is in the bath. Lead also requires to be coated in a bath which is richer in the metals.

In immersing in the bath two different metals, as cast-iron and zinc for instance, the zinc would receive the deposit at once, whilst the iron would not receive the smallest amount of deposit, and in striving to force the metal on the iron surface the operator may impair his solution. Even cast- and wrought-iron require to be coated in different baths. By observing this rule, the solutions are not so liable to get out of order.

Again, iron and zinc require different degrees of battery-power to effect a good deposit upon them. A battery which would coat zinc well would not cause the least deposit to take place upon cast-iron.

Electro-brassing Cast-iron work.—In preparing cast-iron work for the brassing bath, it will be necessary first to make up a "pickle" of the following:—

Sulphuric acid	1 pound.
Water	20 pounds.

The article is placed in the pickle, and allowed to remain until the oxide of iron has become loosened from the surface of the article, in other words, until a brush and sand will easily remove the oxide. If at any time the oxide is found to adhere firmly to the cast-iron surface, the pickling process must be continued until it yields readily to the brush.

When the work is very rusty, it may be first placed in a pickle composed of—

Hydrochloric acid	1 pound.
Water	20 pounds.

and any parts which may have a thick coating of rust may be cleaned by applying strong hydrochloric acid to the part, which readily dissolves the rust. It is better to remove the rust, as suggested, before immersing the whole article in the first pickle. Generally speaking, from an hour to an hour and a half is sufficient time to remove the oxide of iron in the pickling bath.

As soon as the articles are pickled, they are to be well rinsed, and are then to be laid on a board, placed over a vessel of water, called the "cleaning-board," and are to be thoroughly cleaned with a hard brush, sand, and water until the oxide is completely removed. The article is then to be rinsed in clean water, and may be placed in a weak solution of potash or soda. It is now ready for the bath, in which it may be suspended by a stout copper wire connected with the negative electrode of the battery.

For most purposes I prefer using two cells of a Bunsen's battery, consisting each of a cylindrical stone jar A fitted with a cylinder of zinc, C, which must be well amalgamated and a copper wire attached to it. A porous cell B is placed in the centre, and a bar of carbon D is put into the cell, which is then filled with concentrated nitric acid. Into the outer cell is poured a solution of sulphuric acid, consisting of about 1 part of acid to 20 of water. A binding-screw is attached to the carbon, and a stout copper wire, which is to be soldered to a brass anode.

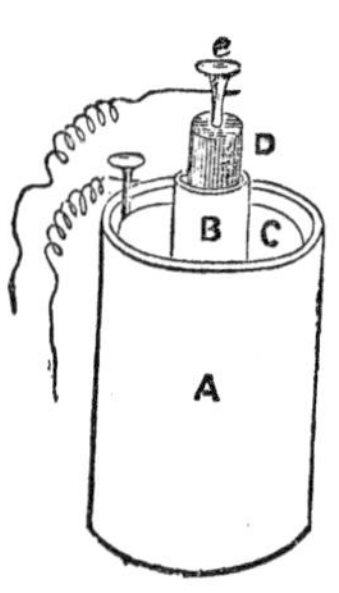

When the article has been immersed in the solution for a few minutes, a white foam will show itself at the point of the wire, in most instances, and frequently bubbles of gas will be seen to rise in various parts of the solution. In electro-brassing, generally, but little deposition takes place unless there is the evidence of chemical action alluded to.

As soon as the article has received the required coating—which, for ordinary purposes, may be accomplished in about two hours, with two cells of the battery just described, holding about four gallons in each jar—it is to be at once rinsed in hot water and then placed in hot saw-dust. For this purpose mahogany saw-dust answers very well. When thoroughly dry, if it is required to bronze it, the article should be rubbed over with a leather and a little powdered pumice or whitening, in order to brighten those surfaces which are to look bright when the work is complete. Instead of bronzing it, the article may be cleaned and lacquered. The bronzing process is described at page 37.

Electro-brassing Wrought Ironwork.—It is more easy to electro-brass wrought than cast ironwork, as it is less porous, and is in general much more smooth. The goods may be first pickled in the sulphuric acid pickle-bath, and then cleaned with a brush, sand, and water. The solution in which wrought ironwork is brassed need not contain quite so much metal as that for cast iron, and it generally does not require the exposure of so great a surface of anode.

When the goods are placed in the bath, if the deposit appears of too red a hue, rather more anode must be exposed; if, on the other hand, the work is pale, less of the anode should be immersed. The surface of anode will generally regulate the colour of the deposit. Wrought iron receives the deposit more readily than cast iron, consequently it need not remain in the bath quite so long as the latter.

Electro-brassing Articles of Zinc.—Goods of this description should first be placed for a quarter of an hour or so in a pickle consisting of:

Sulphuric acid	1 ounce.
Hydrochloric acid	2 ounces.
Water	1 gallon.

The articles may then be rinsed in clean cold water, and, being placed on the cleaning board, they should be scoured well with a hard brush, sand, and water. Zinc goods, if the battery and bath are in good working order, will receive the deposit immediately on being immersed in the bath; when the reverse is the case, however, either the solution is deficient in conducting power—in which case add fresh cyanide and liquid ammonia—the battery is weak, or the surface of anode must be increased. The goods, when sufficiently coated, are to be rinsed in hot water, and then placed in hot mahogany saw-dust; it is important that the articles should be well rinsed. This class of work may be bronzed or polished and lacquered.

Electro-brassing Lead and Pewter Articles.— Lead does not receive the deposit so favourably as zinc, but pewter receives it tolerably well. They may, however, be both coated in the same bath without harm. Lead should be pickled in a dilute solution of nitric acid, say a mixture containing about four ounces of nitric acid to the gallon of water. The same pickle will do for pewter work. The goods may remain in the pickle for half an hour, when they are to be well rinsed and scoured with sand as before; lastly, rinsing in clean water. A good surface of anode should be exposed, more especially when the articles are first put into the bath. If the battery power is not ample, lead is very apt to become coated *in parts*, owing to its being a very indifferent conductor of the current.

In brassing lead and pewter, it is advisable to raise

the temperature of the bath to about 90° Fahr., when this can be done conveniently. The same observation applies to coating tin with brass.

When a brassing solution has been worked for some time, it is liable to deposit either copper or zinc only—generally speaking the former. The principal cause of this is that the solution has not the power of dissolving the brass anode equally, the copper in the alloy being readily attacked by the cyanide and consequently entering the solution, whilst the zinc, being liable to be converted into an almost insoluble salt, either remains on the anode in the form of a white mass, or falls to the bottom of the bath, but a small portion of it entering the solution; consequently the latter, instead of depositing the yellow alloy, deposits only, or chiefly, copper. When the bath is thus found to work unfavourably, it will be necessary to add a little more concentrated solution of zinc; but sometimes I have found that the addition of a large quantity of liquid ammonia to the bath has dissolved a considerable portion of the zinc precipitate from the anode and from the bottom of the vessel, and thus the bath has become restored to its proper condition. When adding the liquid ammonia—which should be the strongest which is made—it is as well to add a little additional cyanide.

The bath may also be restored by separating the clear solution from the precipitate which has fallen to the bottom of the vessel, and by then treating the latter with liquid ammonia and cyanide, which will dissolve the greater portion of it. This is the best plan when it is practicable. The solution thus formed may be added to the bulk of the solution, which will then work well again. Therefore, whenever the anode assumes the white appearance referred to, liquid ammonia or cyanide, or both, must be added, otherwise copper only will be deposited.

Sometimes the bath will become deprived of both zinc and copper, in consequence of the anode not keeping up the supply in the solution. When this is the case, a strong solution of brass should be added to the bath: in fact, a supply of concentrated brass solution should always be kept on hand to be thus employed in case of emergency, for the best solutions are apt to become deprived of metal after being worked a good deal.

Electro-deposition of Platinum.—A solution of platinum may be made by dissolving a piece of the metal in two parts of hydrochloric acid and one part nitric acid, over a sand bath. The acids must both be very strong, or the metal will not yield to their action. When the platinum is dissolved, the acid should be expelled in the same way as that recommended in forming the chloride of gold. A reddish mass will be obtained, which is the chloride of platinum. A little distilled water is now added to dissolve the chloride, into which put a small lump of cyanide, which will at first precipitate and then re-dissolve the platinum. The solution should have about five dwts. of metal to the quart. In working it, the solution should be warm. It is better, however, to filter before using the platinum solution, to remove the impurities with which the cyanide is contaminated.

The battery power employed for depositing platinum should be rather weak, or the metal will be thrown down in the form of a black powder, possessing but little resemblance to the metal itself.

As a platinum anode will not be acted upon by the cyanide, the solution will, of course, soon yield the metal of which it was composed; therefore, it will be necessary, from time to time, to add fresh chloride of platinum to keep it in working order. If it is desired to coat an article strongly with platinum, it will be necessary to keep on adding chloride of platinum to the solution every now and then while deposition is going on, until the object is accomplished. This of course renders the process of electro-platinising not only expensive but tedious; and, for general purposes, impracticable. The cyanide will hold but a small quantity of platinum in solution.

Palladium may be somewhat more readily deposited from its solution than platinum. The metal is to be dissolved in nitro-hydrochloric acid in the same way as above. The solution is next to be treated with cyanide, which will precipitate the metal, and finally re-dissolve it. The solution may be worked warm.

The palladium anode will be acted upon by the cyanide, consequently the operator may deposit the metal to any desirable extent. There is, however, but little importance attached to the deposition of this metal.

Lead.—A solution of lead for the purposes of electro-deposition, may be formed by dissolving the acetate or nitrate of lead in water. By employing the solution in a rather weak state, with moderate battery power, lead may be deposited with ease, but the deposit from these acid solutions is of a very indifferent quality. An alkaline solution may be made by precipitating the lead from either of the above solutions, either with soda, potassa, or ammonia, and then re-dissolving with cyanide, but the solution is only fit for experimental purposes.

Nickel may be easily deposited. Dissolve the metal in nitric acid. Expel the acid by heat, as before. Now add cyanide, which precipitates the nickel, and then re-dissolves it. The solution is now ready for use. An anode of nickel is employed, with moderate battery power.

Or the nickel may be precipitated from its acid solution by potassa, soda, or ammonia. Wash the precipitate and then dissolve with cyanide of potassium.

Iron may be deposited from a solution of the sul phate of iron (copperas), to which a little free sulphuric acid has been added.

This metal may also be precipitated from the solution of sulphate, with an alkali, and again dissolved in cyanide; but the deposition is not of any practical value.

Antimony, Bismuth and Cadmium.—These metals may be deposited from an acid or alkaline solution. If the latter, cyanide is employed as the solvent.

Tin.—Although it is somewhat difficult to form an alkaline solution of this metal, yet this may be done with care; there is but little practical importance, however, attached to the deposition of this metal. Some of the salts of tin are soluble in liquid ammonia, caustic potassa, or caustic soda, and the metal may be deposited from either of these solutions: the addition of a little cyanide seems to favour the rapidity of the deposit.

Tin may also be deposited from an acid solution, the protochloride, for instance; and a very beautiful effect is produced by bringing the anode and cathode within an inch of each other, in which case a fine mass of crystals of tin will start out from the negative pole, approach the positive pole, and gradually assume many beautiful and eccentric forms. The slightest motion causes the crystals to fall from the electrode.

Zinc.—Many persons have tried to deposit this metal from acid solutions—more especially from a solution of the sulphate—but for all practical purposes the processes have been a failure. The principle of depositing metals upon each other in an acid solution is bad, owing to the fact that the metal, when coming in contact with the acid solution, generally becomes acted upon without the aid of the battery.

In 1855 I patented a process for depositing zinc from an alkaline solution, which gave exceedingly beautiful results, and the metal deposited thereby was tough, reguline, and otherwise well suited to many practical purposes. As I believe the process is susceptible of many practical applications, it will be found fully described in the subjoined

SPECIFICATION.

My Invention consists, firstly, in forming a solution for the purpose of coating iron or steel with zinc by galvanic agency. To form the solution I proceed as follows:—I dissolve 200 ounces of commercial cyanide of potassium in twenty gallons of water (rain-water or distilled water being preferable) in a suitable vessel; I then pour into this solution 80 ounces by measure of strong liquid ammonia (of the specific gravity of ·880 I prefer). Having stirred these compounds together, I place several large porous cells, such as those used in Daniell's batteries, in this solution, and pour into each of the porous cells as much of a strong solution of a cyanide of potassium (say, about 16 ounces to the gallon) as will be equal to the height of the solution in the larger vessel; I then attach several pieces of metal, copper or iron by preference, to pieces of copper wire, which are then to be attached to the negative pole of a galvanic battery. These pieces of copper or iron are to be placed in the porous cells. I next attach a piece or several pieces of zinc to the positive pole of the battery, and I then immerse these pieces of zinc in the solution of cyanide of potassium and ammonia. For the above purpose I prefer using good milled zinc, the weight of which is to be ascertained before immersion; and I think it better to "pickle" the zinc slightly, previous to immersion in the cyanide solution, with dilute hydrochloric acid, after which process it should be well rinsed in clean water. The galvanic battery is now to be set in action, and allowed to continue in action on the above materials until the solution of cyanide of potassium and ammonia has taken up about sixty ounces of zinc, that is to say, about three ounces to the gallon of solution.

As soon as the pieces of zinc have been weighed to determine the amount dissolved into the cyanide solution, I dip them into dilute hydrochloric acid, and then rinse them, when they are placed aside for future operations, if necessary; the porous cells are then to be removed. I now dissolve 80 ounces of a carbonated alkali (I prefer the carbonate of potassa) in a portion of the above solution, and when dissolved I add it to the original solution, and stir the whole together for a few moments, after which I allow the solution to stand undisturbed until the sediment formed has subsided; I then transfer the clear solution to another vesse., when it is ready for use.

The above solution may be made in a more concentrated form, say, with half the quantity of water, and it may be diluted down to the required strength by adding more water when wanted to be worked. I prepare cast or wrought iron or steel to be coated with the above solution in the following manner, having first made a pickle composed of,—

Sulphuric acid	1 ponnd.
Hydrochloric acid	½ ,,
Water	2 gallons.

The articles to be coated are first plunged into the above pickle, and allowed to remain until the oxide of iron is easily removable with a brush, sand, and water. As soon as the articles are sufficiently pickled, they are to be rinsed in clean water, and are then to be cleaned with a hard brush, sand, and water; and any oxide which may not have been quite removed by the pickle may be scraped or otherwise removed, or the article be returned to the pickling bath until these parts yield to the brush and sand; or the iron or steel may be cleaned by the processes ordinarily used at the "galvanised iron works." When the articles to be coated are quite free from oxide, they are to be well rinsed in clean water, and immediately placed in the zincing bath, in connection with the negative pole of the battery.

As soon as the articles have received a slight coating they should be removed from the bath and examined, in order to ascertain if there are any parts remaining unclean, in which case those parts should be cleaned, and the whole article once more brushed all over as before and then returned to the bath, where it is to remain until sufficiently coated. It is as well, however, to move the article about in the solution occasionally, in order to insure an uniform deposit. As soon as the articles are sufficiently coated they are to be removed from the bath and rinsed in clean water (hot water being preferable), and they may then be placed in saw-dust to dry them. The articles may be rendered bright either by means of the scratch-brush, or by gently scouring with silver sand, water, and a soft brush. When the above solution has been in use some time it will be necessary to add occasionally a little cyanide of potassium and liquid ammonia, so as to keep the solution at as near as possible the original strength; and if the solution, from being worked with too small a surface of positive electrode, or from other causes has become deprived of a portion of its zinc, I place in the solution (by suspension or otherwise) several porous cells, which I fill with strong solution of cyanide of potassium, and into which I put pieces of copper or iron as before, in connection with the negative pole of the battery, and the zinc electrodes I attach to the positive pole of the battery, by which means I am enabled to keep up the strength of the solution. The above solution is to be worked with zinc electrodes (milled zinc being preferable), and it will be necessary in coating flat surfaces especially to place a piece of zinc on each side of the article to be coated; for instance, if sheets of iron are to be coated, they are to be placed in the solution alternately, that is to say, sheet zinc, sheet iron, sheet zinc, and so on, (the sheets of iron and zinc exposing about the same area of surface,) otherwise the surface opposite the zinc electrode will receive the greatest amount of deposit. I prefer using a battery which yields a considerable quantity of electricity, and the action of which can be maintained for a considerable time without losing its power, by which means I not only secure a good deposit, but uniform results. The battery which I prefer is that known as Bunsen's battery, or a battery composed of carbon and zinc elements. Two or more 4-gallon cells of this battery may be used when large articles are to be coated, or when a considerable quantity of work is to be done in the bath at one time. When cast or wrought iron or steel has become much rusted, it may be cleaned with strong hydrochloric acid, or with a strong pickle composed of hydrochloric acid and water, but it must not be allowed to remain too long in this pickle, or the iron or steel will be acted upon.

Articles of cutlery may be coated by the above process to preserve them from oxidation or rust in damp climates or during sea voyages, &c., and as they will only need a slight coating for this purpose, they will not require to remain long in the bath. In pickling bright steel articles, I should not recommend the use of any hydrochloric acid.

Electro-deposition of Alloys of Metals.—Besides the metals already referred to, I have succeeded in depositing an alloy of copper and nickel, forming a very good quality of German silver, by dissolving German silver in nitric acid, precipitating with an alkali, and re-dissolving with cyanide of potassium.

Silver and gold—forming what jewellers term "green gold"—may also be deposited by adding to a solution of gold a small quantity of solution of silver, but the solution must be worked hot, and with weak battery power. Copper and gold may also be deposited together in the same way. However, the only alloy which seems to have much practical value in it is that of brass, and an extensive manufacture is now being carried on in this art both in England and on the continent.

I trust that the reader will find in the foregoing pages sufficient practical matter to enable him to work the various processes described with facility and certainty of success. I have endeavoured to divest the details of any unnecessary technicality, and to give each process with as much conciseness and simplicity as possible, in order that the student might at once arrive at the readiest mode of setting to work. Those processes which I have dwelt fully upon, are those which I have found most practicable and economical, and consequently more likely to succeed in the hands of the uninitiated.

I propose giving, in the form of an appendix, a few practical memoranda, which I hope may prove equally serviceable to the beginner and the practical electro-metallurgist.

APPENDIX.

1.—In dissolving gold in *aqua regia* (two parts hydrochloric acid and one part nitric) care must be taken that the acids are pure, and that the gold be perfectly *fine*, or at least not inferior to sovereign gold.

2.—If the operator requires to dissolve gold of inferior quality, for the purpose of making his gold solution, he should first treat the gold in the following manner:—To one ounce of alloyed gold of the same quality as that which "colored" gold chains are made of, add two ounces of silver. These are now to be placed in a crucible and melted in a furnace, a little borax being added as a flux. As soon as the alloy is thoroughly melted, it should be poured into a deep vessel of cold water (kept well stirred in a circular direction all the time), and thus it will become "granulated," as it is termed. The granulated metal is now to be removed, and placed in a Florence flask, and to be treated with one part nitric acid and two parts water. This is allowed to digest for an hour or so; applying gentle heat when the chemical action diminishes in vigour. The nitric acid will remove all the copper and silver from the gold, which latter will remain at the bottom of the flask in the form of a dark brown irregular mass. The acid, which will have acquired a green colour, may now be poured into a separate vessel. It will be well to add a little fresh acid to the gold, applying heat as before, in order to be sure that all the copper and silver have been removed. If the acid does not produce chemical action (which may be seen by the absence of red fumes in the body of the flask), the operation is complete. The gold is now to be washed well with hot water, and the washings are to be added to the first solution which was poured from the flask. The gold in its present state may be dissolved with nitro-hydrochloric acid, and thus converted into chloride; or it may be dried, mixed with a little dry potash, and fused in a crucible. When melted, the gold may be granulated, or poured into an ingot.

The solution of nitrate of silver and nitrate of copper formed above, may be thus treated, in order to collect the former:—Put into the vessel containing the green solution a piece of stout sheet copper. In a few moments the silver will begin to deposit itself upon the copper, and by continuing the process for some time—adding a gentle heat, the whole of the silver will eventually become precipitated in the form of minute crystals. In order to ascertain whether all the silver is thrown down, pour a little of the green liquor into a wine glass, and drop in a little hydrochloric acid, which, if any silver be still present in the solution, will form a white precipitate. If, on the contrary, no precipitation takes place, the green solution may be poured off and thrown away. The silver is to be washed several times, to free it from the copper, and when the last washings pass off clear, the silver may be dried and melted, with a little potash, in a crucible; or it may be dissolved in nitric acid and used for making a plating solution.

The copper may be thrown down from the above solution, when the silver is extracted, by immersing in the solution a few pieces of iron, but it is never worth while to do this except for experimental purposes. The solution, however, may be used with the sulphate of copper batteries.

3. Sometimes, when gilding the insides of mugs, tankards, &c., which are richly chased or embossed, it will be found that the hollow parts do not receive the deposit at all, or very partially. When this is the case, the article must be rinsed and well scratch-brushed, and a little more cyanide added to the solution. The anode should be slightly kept in motion and the battery power increased until the hollow surfaces are coated. Frequent scratch-brushing aids the deposit to a great extent, by imparting a slight film of brass to the surface.

4. Silver filagree brooches and articles which have been annealed and cannot be scratched bright, owing to their peculiarity of construction, are frequently troublesome to gild, for the rough surfaces caused by the fire, in the process of annealing, are indifferent conductors

of the current. It will therefore be advisable to scratch-brush the articles as far as practicable, and to add a little more cyanide of potassium to the solution in which this class of work is to be gilt. The article must be constantly moved about, in solution, until coated all over. The battery-power should be brisk.

5. When articles gild a "foxy" colour, as it is termed, this is either owing to the presence of too much cyanide, excess of battery power, or exposure of too large a surface of anode. When this defect shows itself, raise the anode a little and keep the article in motion while in the bath, or remove the anode altogether and move the work about in the solution for a few seconds. This will generally remedy the defect. The power of the current, however, should be diminished, or the anode will become wasted.

6. When a gold solution has been worked for a long time, it becomes contaminated with organic matter, and the deposit is of an inferior colour in consequence. In this case, I have observed that the solution may be restored to good working condition by evaporating it to dryness, and then adding distilled water to re-dissolve it. A little cyanide should then be added, and the solution filtered for use. The heat required to evaporate a solution to dryness does not, as many people suppose, impair the solution, or decompose it; it merely appears to destroy the organic matter and to prevent its influence in the working of the solution.

7. It is sometimes found impossible to make a gold solution work well which has been in use for some years, even evaporation to dryness failing to restore it. It is therefore better and more economical to abandon it altogether and make another. The gold from old solu-

tions may be recovered by means of the battery, or by precipitating the gold with acid. If the former plan is adopted, a piece of copper should be attached to the negative pole of the battery, and another piece (as an anode) be attached to the positive pole. When the battery has been in action for some time, the gold—or at all events the greater part of it—will be deposited upon the negative pole, from which it may be removed by mechanical means, or by dissolving it off with nitro-hydrochloric acid. If it is preferred to throw down the gold from the solution with acid, the solution must be placed in a large vessel *in the open air*, as the fumes which will arise are highly deleterious if breathed, and sulphuric acid poured in carefully until no further effervescence takes place. The precipitate formed should be allowed to subside, when the clear liquor may be poured off and thrown away. The precipitate may be washed with hot water; after which it may be dried, mixed with a little potash, and fused in a crucible until the gold is gathered into a button. The operator will seldom find that he can recover nearly the amount of gold that he put into the solution; owing to the irregularities of working, the solution becomes deprived of a considerable proportion of gold, and I have frequently found that old solutions will yield scarcely any metal worth speaking of.

8. In gilding, if a copper and a silver article be immersed in the solution together, the copper article will receive the deposit first, and the silver article will be troublesome to gild sometimes under such circumstances; and in trying to force the gold upon the silver, probably the copper article will receive the deposit so quickly that it will be liable to strip off when scratch-

brushed. The silver article, therefore, should be placed in the solution first, and when it is coated, the copper one may be suspended by its side.

9. Each metal to be gilt plated or brassed, should have a solution for itself, otherwise the bath in which several different metals have been coated, will become impaired, unless, however, each metal has first been coated by itself to some extent.

10. When it is found that the operator cannot, from some cause or other, produce a good colour when gilding, it is useful to have at command the means of improving the colour without the trouble and annoyance of persevering with an indifferent gold solution. Some gilders employ the following mixture to give an artificial colour to gilt work; and provided the work is strongly coated, it may be used with advantage:—

Alum	3 ounces.
Nitrate of potassa (saltpetre)	6 ,,
Sulphate of zinc	3 ,,
Common salt	3 ,,

Mix the above materials into the form of a thick paste, dip the articles in it, or brush them over with the compound, and place them on a piece of sheet iron. The iron is to be heated over a clear charcoal or coke fire, until the articles appear nearly black, when they are to be plunged into cold water. A very useful formula, and one which may be used with less care than the above—especially for small work, is the following:—

	OZS.	DWTS.	GRS.
Sulphate of copper	0	2	0
French verdigris	0	4	12
Chloride of ammonium (Sal ammoniac)	0	4	0
Nitrate of potassa	0	4	0
Acetic acid (about)	1	0	0

Reduce the sulphate of copper, sal-ammoniac and nitrate of potassa to a powder, in a mortar; then add the verdigris, and pour in, little by little, the acetic acid, stirring well all the time ; the whole will assume a bluish-green mass. The article to be coloured is to be dipped into this mixture, and, being placed on a piece of sheet copper, heat is to be applied until it assumes a black colour. It is allowed to cool, and is then placed in a tolerably strong sulphuric acid pickle, which will dissolve the colouring salts, and the article will assume a rich fine gold colour. It is sometimes advantageous to scratch-brush the article before submitting it to the above process, and it will then come out of the pickle perfectly bright. The article, when removed from the pickle, is to be well rinsed in warm water to which a little potash has been added. A soft brush and soap and warm water, skilfully applied, will tend much to improve the article—especially if the work is either chased or embossed.

11. Moving the articles about in the bath, will at all times enable the operator to vary the colour of the deposit from pale straw-colour to a very dark red. The temperature of the solution likewise influences the colour of the deposit, the colour being lightest when the solution is cold, and gradually becoming darker as the temperature increases. Variations in the surface of anode exposed while the articles are in solution, will also alter the colour of the deposit. The amount of cyanide in the bath and the strength of battery-power, influence the deposit in the same way.

12. If there be not sufficient cyanide in the gold solution, the anode will not become freely dissolved; consequently, as has been shown, the solution will soon become exhausted of its gold, and the articles gild of an inferior colour. Adding more cyanide, under such circumstances, will not remedy the defect, but a little concentrated solution of gold should also be added at the same time.

13. In gilding watch-movements, the greatest care must be observed with regard to cleanliness. The work is first to be placed in a weak solution of caustic potassa for a few minutes, and then rinsed in cold water. The movements are now to be dipped in pickling acid (nitrous acid) for *an instant*, and then plunged *immediately* into cold water. After being finally rinsed in hot water, they may be placed in the gilding-bath and allowed to remain until they have received the required coating. A few seconds will generally be sufficient, as this class of work does not require to be very strongly gilt. When gilt, the movements are to be rinsed in warm water, and scratch-brushed; they may then be returned to the bath, for an instant, to give them a good colour. Lastly, rinse in hot water, and place the movements in clean box sawdust. An economical mode of gilding watch-movements, is to employ a copper anode—working from the solution—which must be re-supplied with gold from time to time as the solution becomes exhausted.

14. When an article is immersed in the silver solution, if it assumes a dark colour, either the solution is too rich in cyanide, the battery-power is excessive, or too large a surface of anode is employed. Any one, or all of these conditions combined, will cause this defect. The operator should at once remove the article (unless it be made of Britannia-metal, pewter, or lead), and have it cleaned again by the usual process. It may

then be returned to the bath, and a much smaller surface of anode exposed. This will at once alter the colour of the deposit, and the anode can be lowered a little from time to time, to increase the speed of the operation. Should the article, however, still receive a dark-coloured deposit, either the solution must be weakened with water, or the battery-power reduced. But the solution should not be altered until the other remedies have been tried.

15. When it is desired to give to a plated article, or a portion of the same, that appearance which is technically termed "oxidation," any of the following processes may be employed with success. Sometimes very pleasing effects may be produced upon silver work by the "oxidising" proceses.

1. Dissolve 1 dwt. of platinum in *aqua regia*. Evaporate the acid, and when the resulting red mass is quite cold, dissolve in a little sulphuric ether or alcohol. Or the chloride of platinum may be dissolved in cold water, or used, in its acid state, before evaporation. Apply with a camel's hair pencil to those parts which are required to be "oxidised," and as soon as the spirit or ether has evaporated, the pellicle of platinum remaining will give the appearance required.

2.	Sulphate of copper	2 dwts.
	Nitrate of potassa	1 ,,
	Muriate of ammonia	2 ,,

Dissolve in a little acetic acid. Apply with a camel's hair pencil. The article should be warmed before using this mixture.

3. Hydrosulphate of ammonia, strong or diluted, will give either a dark or light tint of oxidation.

4. The fumes of sulphur will give to silver an extremely beautiful blue steel-like surface. The operation should be conducted in a closed box, all parts of the article not to be coloured being protected with a suitable cement or wax.

5. Nitric acid alone will produce an oxidised surface upon silver.

16. Certain parts of an ornamental silver article may have a very pleasing effect produced upon them by oxidising some parts, gilding others, and then depositing a slight coating of copper upon small portions of the article, which may be done in the following manner:—Dissolve a little sulphate of copper, and add a few drops of sulphuric acid; apply this solution to the part to be coated, with a camel hair brush, now touch the moistened part with a piece of steel wire, and it will instantly become coated with copper. Any design can be worked in copper by this means; but it is not necessary to state, that the amount of copper deposited is very trifling, consequently the article should not be subjected to much wear.

17. When a silver solution works badly, and it appears impossible to restore it by the ordinary means, the operator may precipitate the silver with sulphuric acid in the manner described for gold (page 96). The precipitate is to be well washed, and may then be re-dissolved with cyanide of potassium. Water is then to be added to make the required amount of solution, and in all probability it will work as well as ever. The solution should be filtered before using, which may be conveniently done before adding the bulk of water.

18. If, when silver anodes are used, the solution contains a great excess of free cyanide, the anodes will

dissolve away irregularly, and sometimes numerous small particles of silver drop off the anode; these particles, if allowed to fall on the work to be plated, will render it rough. It is therefore, under such circumstances, advisable to place the anode in a canvas bag, or a bag made of Holland linen, by which means the small granules of silver will be retained, and may be collected from time to time, and melted or dissolved to make nitrate of silver.

19. In "stripping" articles, by the process given at page 56, the operator must take care that the fumes arising from the process are not allowed to enter the apartment in which he is operating, as they are exceedingly offensive and injurious. The process may be carried on upon a sand-bath, with a flue above, or upon the hob of an ordinary stove.

20. Sometimes the silver thrown down from stripping solutions, when melted and cast into an ingot, will not, if submitted to the "flatting-mill" to be rolled out, roll well, but will crack under the operation. This defect is probably owing to the presence of a small quantity of zinc. It is better, in such a case, to remelt the silver with a small portion of copper, or to throw in some nitrate of potassa when the silver is in a state of fusion.

21. When silver or electro-plated articles have become much tarnished by exposure to the atmosphere, the surfaces may be cleaned by brushing over them a strong solution of cyanide of potassium; or strong liquid ammonia will answer the same purpose. Jewellers' rouge, in the form of a paste, applied with a stiff brush, will render the surface of chased work clean, but the bright surfaces should be polished with the

palm of the hand, and moist rouge, which is rubbed on until it becomes quite dry, and the hand appears black from the silver which has worked off the article by friction. When the articles are very dull, a little rotten-stone may be applied before the rouging process.

22. Gold may be stripped from articles which have been gilt, by placing them in strong nitric acid, to which a little dry common salt has been added. When the gold is removed, the articles should be at once rinsed and cleaned in the ordinary way. When the nitric acid stripping solution has been worked a good deal it removes the gold but tardily; it should then be cast aside, and the gold collected by evaporating the solution to dryness, and fusing the residuum with a little potassa or soda. When the gold is fused into a button, a little nitre may be added occasionally, in order to refine it thoroughly.

23. As deposition takes place more rapidly upon those surfaces which are nearest the anode, it will be necessary, in order to coat the goods as uniformly as possible, to move them occasionally, and to present a different surface to the anode. By doing this frequently, a tolerably uniform deposit may be secured. Or the anode may be shifted to effect the same result. Presuming that electro-deposition takes place principally where the article is in "electrical sight" with the anode, it is well to surround the work with surfaces of metal to be dissolved, which will save the necessity of frequently moving the articles in solution.

The insides of cream-ewers, sugar-bowls, teapots, &c., since they cannot be placed in such a position that the insides will be exactly opposite the anode, may first receive a deposit inside by filling the vessel with silver solution, and proceeding in the same way as in gilding the inside of a vessel. Or this may be left until the article is well coated outside. Generally speaking, it is better to let the solution be slightly moved occasionally, in order to expose fresh surfaces of the solution to the work being plated.

24. Stout copper wires conduct the current better than fine wires; consequently, the wire employed in connecting the work to be plated and the anodes with the battery should be always sufficiently stout to carry the current freely. For a single cell, wire about one-sixteenth of an inch in thickness will be sufficient, but where a series of cells are employed, in electro-brassing for instance, the wire should be much thicker,—say at least one-eighth of an inch thick. If too thin a wire is used, when the battery is in circuit,—that is, when it is connected with the anode and goods to be coated,—the wire will sometimes become quite hot, owing to its being unable to convey the amount of current generated.

25. It is always advisable to commence the process of electro-deposition with moderate battery power, which may be augmented after a little while, except in those cases referred to in a former part of this work. If too strong a current is employed, the articles will, in all probability, "strip" in the process of scratch-brushing or burnishing. Again, when the battery power is very strong, the solution becomes decomposed. As a rule, no effervescence or frothing should be allowed to take place in a plating bath.

26. Electro-deposition will proceed much quicker when the temperature of the atmosphere or solution is high; therefore, the operator should observe that the surface of anode be not too great at first, or the battery power excessive, or the deposit will be faulty. Whenever a warm solution is employed, deposition should be allowed to commence slowly at first, gradually lowering the anode as the coat thickens.

27. Deposition from silver or gold solutions takes place more actively upon brass, German silver, or copper, than upon silver or gold; therefore the operation must be carried on slowly at first. When a copper or brass article becomes covered with silver, deposition does not proceed quite so rapidly as at first, when the inferior metal was exposed in the solution; the manipulator may therefore accelerate the speed of the operation as before recommended. The same operation applies to gilding. Gold is more easily deposited upon brass or copper than upon gold; therefore, after the first layer has been deposited, the operation may be carried on more vigorously.

28. The apartment in which electro-deposition is carried on should be kept as dry as possible, and the temperature at about 60° F. In warm weather, when the apartment assumes a higher temperature, the strength of battery power, &c., should be regulated accordingly, otherwise deposition will take place too rapidly.

29. Batteries employed for this purpose, should be made to work as uniformly as possible. When a battery works slowly, it is better to take on an extra cell of about equal power, than to make frequent additions of acid to the battery, which is apt to cause it to act irregularly.

30. The zinc employed in a battery, when not excited by salt and water, should be "amalgamated." This

is accomplished by placing some mercury in a dish with a little hydrochloric acid. A piece of flannel or baize, tied to the end of a stick, and dipped into the acid and mercury, is to be rubbed all over the cylinder or plate of zinc, until it assumes the characteristic brightness of mercury. When a cylinder of zinc is to be amalgamated, I have found that putting mercury into a coarse flannel bag, dipped now and then into hydrochloric acid, and applying it first to the outside of the cylinder, renders the process of amalgamating this surface very simple and effective, when only a small quantity of mercury is at hand. This is a very economical method, as with care little or no waste of mercury occurs. When the amalgamated plate or cylinder has been in work some time, the operator should observe if "local action" is taking place upon any part of the metal. When this is the case, it is accompanied by a violent effervescence within the cell. The cylinder, &c., should be at once removed from the cell, and those parts which have been most violently attacked by the acid solution, must be re-amalgamated. Where local action takes place, the part is generally of a dull and dark grey colour.

31. The copper cylinders and plates used in batteries, should be occasionally cleaned with dipping nitrous acid, and then rinsed in cold water, or they may be scoured with sand and a hard brush.

32. When the copper conducting wires become corroded by being splashed with solution, &c., they should be cleaned with a piece of emery cloth. I have found it advantageous to coat these wires with silver—this metal being less liable to corrosion than copper.

33. In using binding screws for connecting the wires

of a battery with the anode and goods to be coated, the operator must take care that the connection between the point of the binding screw and the wire is clean, otherwise the current will be conveyed partially or not at all. It is well to slightly file the point of the screw before using it, or to roughen it with a piece of coarse emery cloth, which will enable it to *grip* the wire better. The hole through which the wire is passed, may be kept clean with a small round file. When binding screws become very much corroded, they may be pickled in sulphuric acid and water, for a few hours, and then cleaned with a hard brush and sand; or they may be dipped in nitrous acid.

34. Strips of sheet copper will be found very convenient substitutes for copper conducting wires. The strips may be cut from a sheet about $\frac{1}{32}$ of an inch in thickness.

35. In working a battery for electro-deposition, the operator must secure a considerable *quantity* of electricity of sufficient *intensity* to give the necessary activity to that quantity. I prefer, when arranging two or more cells of a battery, to deposit silver, attaching the wires connected with the zinc elements to the metal rod on which the articles to be coated are suspended, and the wires proceeding from the copper elements I attach to the anodes. By this arrangement, the quantity is multiplied by the number of cells employed; whereas,

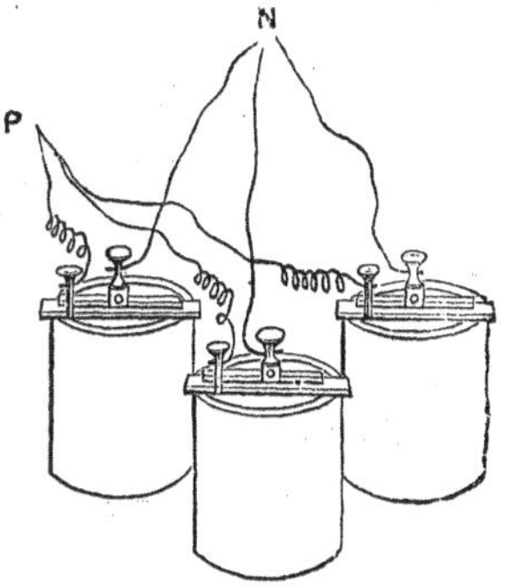

if the cells are alternated—that is, the zinc of the one cell to the copper of the next, and so on—the intensity is multiplied and the arrangement only gives the quantity of one cell.

36. When it is desirable to deposit the metal in a *hard* state, it may be advantageous to alternate the cells, so as to increase the *intensity* of the current, as this quality of current seems to affect the nature of the deposit. A good tough reguline deposit appears to be dependent upon the current being feeble in intensity, but considerable in quantity.

37. In depositing brass, however, the reverse seems to be the case, for here greater intensity is absolutely necessary, or the copper alone will be deposited. I have generally found that at least two cells of the zinc and carbon (an intense) battery, alternately arranged, have been necessary to obtain a deposit of good colour. Again, if the battery be too powerful, the zinc only will be deposited. The exact *mean* appears to be absolutely necessary to obtain good results in electro-brassing.

38. When the anodes are only partially immersed in the solution, and have been worked for some time, the metal will dissolve off rapidly at the surface which is just out of the solution, and probably the anode may be divided, and fall into the solution. It is advisable, therefore, occasionally to shift the position of the anode in order to prevent this local action upon it. It is a good plan to suspend the anodes employed in gilding by stout platinum wires, so that the whole of the anode may be immersed when necessary, without the solution being injured.

39. In preparing solutions, more especially gold and silver solutions, distilled water should be employed;

but, when large quantities of solution are required, this may be inconvenient; therefore, rain-water, if it can be obtained, may be substituted; or water which has been boiled, losing some of its impurities, may be used in preference to common water. Pump-water is very objectionable. If rain-water is employed, it should be filtered before using; and it is better to collect it as 't falls *direct* from the atmosphere, rather than use that which falls off the roof of a house, &c.

40. Electro-deposition of gold and silver may be carried on by the "single cell" arrangement; but, although very good results may be obtained by it, it is of very little commercial importance. The operations of gilding and plating, when conducted by the separate battery, are so simple that even the "single cell" process, simple as it is, will scarcely be employed, except for experiment.

41. In gilding or plating by the "single cell" process, however, a jar is fitted with a cylinder of zinc inside, which is excited either with sulphuric acid and water, or salt and water. A porous cell is placed in the centre, which is filled with either gold or silver solution. A strong copper wire is soldered to the zinc, to which the article to be gilt or plated is suspended, by means of a thinner wire, and the moment the article is immersed deposition takes place.

42. It is advisable to *anneal* the anodes before using them. This may readily be done by making them red hot over a clear fire (a charcoal fire being preferable), and then allowing them to cool. The anodes of gold, silver, copper, and brass may be plunged into dilute sulphuric acid after they have been annealed, by which their surfaces will be rendered quite clean and free from

the "fire mark." Brass and copper anodes may be dipped in nitrous acid for a moment, and then plunged into cold water.

43. Cyanide of potassium may be prepared, for electro-chemical purposes, by the following process:— A quantity of commercial ferrocyanide of potassium is to be reduced to a powder; it is then to be roasted on an iron slab, or piece of sheet-iron with its edges turned up to prevent the material falling off. The heat is to be continued until the substance is quite free from water of crystallisation, which will become evident by its losing its transparency. If the heat be applied too suddenly, the ferrocyanide is apt to *decrepitate,* and much of it may be lost. Care must also be taken not to apply too much heat, or it will become fused to the iron slab. When the ferrocyanide is dried, it is to be mixed intimately with *dry* carbonate of potassa, in the following proportions:—

Dried ferrocyanide	16 ounces.
,, carbonate of potassa	8 ,,

Both materials being well mixed, they are to be placed in an iron crucible or ladle, which should be previously made hot, and the whole subjected to a strong heat in a coke fire; the heat may be increased as fusion progresses. When the substances have fused into a liquid, they are to remain in this state for about a quarter of an hour; the crucible is then to be removed from the furnace, and its contents allowed to settle for a few moments; the clear liquid may then be carefully poured out, either into a shallow iron mould or upon an iron slab or dry flag-stone. The sediment remaining at the bottom of the crucible should be shaken out while hot, or it will be troublesome to remove it. It is a good plan, while the cyanide is fusing, to dip an iron rod into the mass occasionally, and then to examine the portion thus removed, which will be brown at first, and subsequently white when the process has been carried far enough.

44. Sometimes electro platers have employed ferro-cyanide of potassium (yellow prussiate of potassa) instead of the cyanide in forming silver solutions, but this substance has not been found to answer well, since it has not the power of dissolving the anode, therefore the solution soon becomes exhausted of its silver. Again, it requires so large a quantity of the ferro-cyanide to keep the solution in action, that eventually it crystallises upon the inner surface of the bath.

45. Hyposulphite of soda has also been employed as a substitute for the cyanide of potassium; but since the solution which is formed with it is very readily acted upon by light, it is never likely to become much employed; besides, the solutions made with cyanide of potassium are found, for all practicable purposes, infinitely superior to those made with this or any other substitute.

46. Mr. George Knight, the eminent philosophical instrument maker of Foster Lane, Cheapside, has lately drawn my attention to a very ingenious and portable battery, consisting of two zinc plates and a plate of platinised carbon, or graphite, united by suitable binding screws, and placed in a glass cell. This arrangement, I believe, is due to Mr. C. V. Walker. The battery is excited with sulphuric acid and water, and is capable of yielding a powerful current without any unnecessary waste of material. This battery is certainly one of the

most simple and effective which has been introduced for some years, and in the hands of the student it cannot fail to give many pleasing results with but little trouble and expense.

There are many other processes for the electro-deposition of metals upon each other besides those enumerated in this work; but the author,—being desirous merely to enter into the consideration of those processes which are of a practical nature, and which may be pursued by the student for commercial purposes,—has confined himself, as far as possible, to detailing those processes which are most likely to succeed with the beginner. At the same time it is hoped that many who are now practising the art of electro-metallurgy, either as gilders, platers, or bronzers, may glean from these pages some useful hints, and which the author trusts will repay the reader for the time devoted to the perusal of this little work.

TABLE OF WEIGHTS AND MEASURES.

APOTHECARIES' WEIGHT.

1 Pound	*equals*	12 ounces.
1 Ounce	,,	8 drachms (480 grains*).
1 Drachm	,,	3 scruples.
1 Scruple	,,	20 grains.

TROY WEIGHT.

1 Pound	*equals*	12 ounces.
1 Ounce	,,	20 pennyweights (dwts.) (480 grains*).
1 Pennyweight	,,	24 grains.

IMPERIAL MEASURE.

1 Gallon	*equals*	8 pints.
1 Pint	,,	20 ounces.
1 Ounce	,,	8 drachms
1 Drachm	,,	60 minims.

* An ounce Avoirdupois is only 437·5 grains.

47. Silver articles which are required to be left a dead white should be thus treated :—The article is first to be heated to "cherry redness" (*i.e.*, a dull red heat), and is then to be allowed to cool. When quite cold it is to be placed in a pickle of *very* dilute sulphuric acid (about 5 parts acid in 100 parts water). The article should remain in this pickle for an hour or two; and if not sufficiently uniform in surface, it should be rinsed in hot water, dried spontaneously, and again heated as before—the operation of pickling being repeated. As the object in this process is to remove the copper with which the silver is alloyed, the repeated pickling will not produce any injurious result, but will merely remove the copper from the *surface* of the article, leaving fine silver alone upon the surface. A solution of alum in water may be substituted for dilute sulphuric acid, if preferred. In either case the solution should not be used more than once or twice, and then be thrown away.

After the article has been *whitened* as above, it must be removed from the pickle, and be well rinsed in hot water, which for this purpose must be absolutely clean, and then placed in warm box-sawdust. It is well to keep perfectly clean box-dust for articles which have to be whitened, as the slightest stain in some cases would be fatal to the object sought. In whitening silver watch-dials, great care must be taken not to warp the dial in the process of annealing. The dial should be placed on a perfectly flat piece of charcoal,* face upward, and a gentle blast with the blowpipe carefully applied, and as far as practicable the flame should play *all over* the surface of the dial without

* A flat surface may be given to charcoal by rubbing it upon a flag stone.

absolutely touching it, by which means the dial will soon become sufficiently heated, without becoming warped or in any way injured. Silver dials may also be annealed by placing them upon a flat sheet of copper, which is then brought in contact with a clear fire until red-hot.

48. **Brass** time-piece dials may be whitened, *i.e.*, silvered, by rubbing a mixture of chloride of silver (silver precipitated from the nitrate with common salt or hydrochloric acid) and common salt. The mixture should be worked up into a thinnish paste, and be applied with a soft cork or piece of washleather. The dial is then to be rinsed and placed in box-dust.

49. The sediment which accumulates in the scratch-brush box should be carefully preserved and dried, and this, with other waste of a similar description, collected and fused with dry carbonate of potash; and as the resulting "button" will be an alloy of gold, silver, copper, &c., it will be necessary, in order to refine it, to proceed as follows:—Remelt the alloy, and "granulate" as before described; then place the grains in dilute nitric acid (2 parts acid to 1 part water), which will dissolve all but the gold, the latter remaining as a brown powder at the bottom of the vessel in which the operation is conducted. The solution formed should next be poured off into a jar or basin, and a piece of copper immersed, which will at once throw down all the silver. The silver and gold thus obtained will require to be well washed with hot water, and finally dried and fused.

50. **Gold** articles may be "coloured," as it is termed, by immersing them in the mixture described at page 97, line 17. The mixture should be placed in

a common pipkin, and allowed to fuse, the articles being removed occasionally to ascertain if they be of good colour. After the articles have been removed from the pipkin, they should be allowed to cool, and then immersed in dilute sulphuric or acetic acid, which will remove the flux. When this is done, the articles may be rinsed in a weak solution of potash or soda, and finally brushed with hot soap-and-water; they must then be rinsed in hot water, and placed in clean warm box-saw-dust. A badger-hair brush is the best to remove all traces of sawdust from articles which have been dried in it.

51. In gilding cast brass articles, which are required to be left dead in the hollow and chased surfaces, the best plan is first to wash the article in a solution of caustic soda (a solution of soapmaker's "soda ash" will do) or potash, then, after well rinsing, dip the article *for an instant* in fuming nitric acid, and after it has become thoroughly acted upon all over plunge it *instantly* in cold water: the article should then be well rinsed in hot water, and is then ready for the gilding bath. Articles of this description seldom require to be strongly gilt, the *colour* being the principal object. The *bright* surfaces may require to be burnished.

52. Electro-gilders and platers will do well to keep several burnishing tools on hand for small things which may be needed in a hurry, or which would not be within the ordinary province of a professional burnisher to accomplish. As mere friction is required, the operator may easily acquire the proper knack of using the tools. A small steel burnisher, about twice the size of a ladies' "stiletto" or eyelet-hole piercer, is a convenient tool for edges of brooch mounts, &c.

53. As it may be necessary that the electro-plater

should be able to solder an article at a moment's notice rather than send it out to be done, he may derive advantage from the following hints:—The operator is to be provided with an ordinary "soldering iron" (which is made of copper, by the way!), and the face must be filed smooth with a keen file; after which it should be placed in a fire until hot, but not *red* hot. The soldering iron should then be "tinned," as it is termed; that is, rubbed on a piece of sheet tin with a little rosin and soft solder. As the face of the tool will have become slightly oxidised after removal from the fire, it will be necessary to pass the file over it again in order to clean the surface to be tinned. The moment this is done, plunge it *at once* into the rosin and solder, and by gently working the soldering iron over the surface of sheet tin, it will become coated with solder and fit for use. The soldering iron should never be allowed to be heated to redness, otherwise the solder becomes "burned," as it is termed, and, uniting with the copper, forms a hard alloy which the file will scarcely touch unless the tool is very hot. In soldering an article, the first thing to do is to clean the part to be soldered, which is generally done by scraping the surface with a sharp instrument, such as the point of a penknife, or, still better, a three-square scraper made out of an ordinary three-square file, and great care is necessary to ensure all parts which are to receive the solder being perfectly clean. The operator should provide himself with a solution of chloride of zinc, which is made by dissolving a few pieces of zinc (say half an ounce) in an ounce of hydrochloric acid; this solution may be kept in a wide-mouthed bottle ready for use. The solution of chloride of zinc is applied to the parts to be soldered with a camel-hair brush or the feather of a quill. As soon as this is done, and the parts to be soldered are brought together, the soldering iron should be again heated until it is hot enough to melt the solder freely. It should not only do this, but also take up a *globule* of the melted solder and hold it in suspension until borne by the operator to the part to be soldered. The instant the soldering iron touches the surface to which the chloride of zinc has been applied, the solder will "run" and attach itself readily. It will be necessary to hold a strip of solder in the hand, in order to apply more as required. Rosin may sometimes be employed instead of the chloride of zinc, but as a rule the latter will be found preferable. In connecting copper wire with zinc plates, the solution of chloride of zinc may advantageously contain an excess of acid; in fact, hydrochloric acid alone may be employed in the same way as the chloride of zinc. Small articles may be united with soft solder by the aid of the blowpipe. In this case the solder should be hammered flat, and, after being scraped clean, cut into small pellets; that is, first cutting the solder into strips and then cross-cutting into small squares. These pellets are to be placed upon the article to be soldered (after the parts have been scraped as before, and the chloride of zinc applied), and a jet of flame gently blown upon the article with the blowpipe will readily unite the parts.

54. All goods which are to be plated or gilt should be placed in the bath as soon as possible after being cleaned or scratch-brushed and rinsed: if they are allowed to remain long in water, or exposed to the air, a film of oxide is formed upon the surface of the metals, and which, however slight, has a tendency to prevent the gold or silver adhering firmly to the metal to be coated. It is not a good plan to prepare too many articles at one time for the bath, allowing them to remain in water until a quantity is ready: it is better to get them into the bath as quickly as possible, a few at a time, of course taking care that the surface of anode exposed and the battery power are regulated according to the *surface* of goods immersed in the bath at one time. When the number of things in a bath is increased, the anode must be lowered and the battery power rendered more vigorous. As we have said before, silver and gold do not receive the deposit of these metals so freely as copper, brass, German silver, &c.; therefore, when these metals have become coated to a certain extent with the superior metals, the battery power may be judiciously augmented and the surface of anode increased.

55. A very useful solution of silver or gold for plating or gilding without the aid of a battery may be made as follows:—

Take, say, 1 ounce of nitrate of silver dissolved in 1 quart of distilled or rain water. When thoroughly dissolved, throw in a few crystals of hyposulphite of soda, which will at first form a brown precipitate, but which eventually becomes redissolved if sufficient hyposulphite has been employed. A slight excess of this salt must, however, be added. The solution thus formed may be used for coating small articles of steel, brass, or German silver, by simply dipping a sponge in the solution and rubbing it over the surface of the article to be coated. I have succeeded in coating steel very satisfactorily by this means, and have found the silver so firmly attached to the steel (when the solution has been carefully made) that it has been removed with considerable difficulty. A solution of gold may be made in the same way, and applied as described. A concentrated solution of either gold or silver thus made may be used for coating parts of articles which have stripped or blistered, by applying it with a camel-hair pencil to the part, and touching the spot at the same time with a thin clean strip of zinc.

56. To the practical electro-plater, and even the amateur, it may be useful to become acquainted with the art of "hard soldering," as it is termed; and as there is frequently, especially in small provincial towns, a difficulty in getting even a small job executed with dispatch, we will give the reader a few hints upon the manipulation of this process.

"Hard soldering" consists in uniting any two metals, or parts of the same metal, by means of an alloy composed of two parts of silver to one part of brass. The silver and brass should be melted together as follows:—Having obtained a broad piece of good charcoal, scoop out a slight hollow on the flattest surface to receive the alloy. Now place the metals in the hollow, and fuse them by means of a blowpipe, using either a jet of gas or an oil lamp with a good broad wick. As soon as the metals become hot, touch them with a crystal of borax (borate of soda), which will immediately fuse, and act as a flux. The jet of flame must now be vigorously employed until the metals are completely fused. The fusion may be continued for a few moments in order to ensure perfect amalgamation. When the "button" of solder is well melted, the flat surface of a hammer may be placed quickly upon it, by which means it will become flattened; in this form it may be readily beaten out (unless a pair of steel rollers are at hand) until sufficiently thin to cut with a pair of jewellers' shears. The solder can be hammered or beaten out upon any solid iron surface; but, as each time the blow is given the alloy becomes harder, it will be necessary from time to time to *anneal* it, *i.e.*, place it again upon the charcoal, and apply the blowpipe flame until the alloy is of a "cherry-red" heat; it must then be plunged into cold water, and is ready for beating out or rolling, as the case may be. The object being to make the solder as thin as ordinary card, or even thinner, when the operator is without a pair of rollers he must use the next best substitutes—a hammer and patience. The solder, before being used, must be scraped with a keen steel edge, and then partly cut into thin strips, and these again cross-cut into small pieces or pellets about one-sixteenth of an inch square. These pellets may be cut when required for use, or kept in a clean box used for the purpose. The operator should next provide himself with a clean piece of slate, say about three inches square, and a small phial filled with water, and having a cork with a small groove cut in it from end to end. The bottle is used to apply moisture a drop at a time, whilst a large crystal of borax is rubbed upon the slate. By this means a thick creamy paste of borax is obtained upon the slate, which will be used as directed presently. The parts to be united or soldered must now be scraped clean *wherever the solder is expected to adhere*, and, with a camel-hair brush or quill feather dipped in the borax paste, brush over the parts to be soldered. A few pellets of the solder may be placed on the dry corner of the slate, and with the extreme point of the brush moistened by the paste one pellet at a time may be readily taken up, and placed upon the prepared surface of the article. The article should be placed upon a flat piece of charcoal (made flat by rubbing on a flag-stone), and, if necessary, tied to it by thin "binding wire." A gentle blast of the blowpipe will at first dry the borax, and the flame must then be increased (holding the blowpipe some distance from the flame, in order to give a broad jet), and in a few moments, if the jet is favourable, the solder will "run," as it is termed, into every crevice, and the blowpipe must be *instantly* withdrawn. A very little practice will make the operator expert in this interesting art, and it will be advisable for him to practise upon articles of little value until he has not only acquired the use of the blowpipe, but also the proper kind of flame to make the solder run freely. After an article has been hard soldered, it is allowed to cool, or may be at once placed in a weak solution of sulphuric acid (a few drops of acid to an ounce of water), which, after a few moments, will dissolve the borax flux which remains after the soldering is complete. The article should now be rinsed in cold water and dried.

In carrying out the above operation, it would be well to be provided with everything necessary for the purpose (and, in fact, this should always be the first consideration of the student in practising a new art), as the absence of any requisite would not only entail disappointment, but failure.

57. When a zinc plate is imperfectly amalgamated, local action will set in, and the zinc element becomes powerfully acted upon by the acid employed in the battery. When this is the case, there is evolution of gas from the battery cell. The zinc plate must be withdrawn and re-amalgamated, the part most acted upon being particularly attended to. When a single zinc plate and two copper plates are used, this local action will take place if the copper and zinc elements are too close to each other, in which case the zinc plate generally becomes most freely dissolved towards the centre, but at the upper part of the plate.

58. In order to prevent the zinc plate being *cut off* or dissolved at that part which is between the atmosphere and its contact with the acid solution, it is advisable occasionally to add a little more water to the battery, so as to bring the solution higher up the plate. It is also a good plan to raise the plates a few inches out of the bath occasionally, or to stir the acid solution gently with a stick.

59. When there is a want of activity in the battery (the acid solution being in good order), it may be as well to look to the connections. The binding-screws may require cleaning (a smooth file being used for the purpose, or a piece of emery-cloth wrapped round a flat piece of wood), and the ends of the wire should be well rubbed with emery-cloth.

60. When certain parts of an article require to be gilt whilst others are to be left silvered, it will be necessary to apply certain preparations to the parts to be protected. For this purpose many preparations are used. A solution of gum copal or mastic may be applied with a camel-hair brush; but, unless these varnishes are tolerably thick, they are apt to run. The composition described at page 59 may be used for this purpose, but it will not be safe to use it in hot solutions.

In some cases a strong solution of shellac in alcohol may be used to cover certain parts of an article which require protection.

61. In gilding chains, brooches, pins, rings, and other articles which have been repaired, *i.e.*, hard soldered, sometimes it is found that the gold will not deposit freely upon the soldered parts; when such is the case, a little extra scratch-brushing applied to the part will assist the operation greatly, and I have sometimes found that *dry* scratch-brushing for an instant—that is, without the stream of beer usually employed—renders the surface a better and more uniform conductor, and consequently it will more readily receive the deposit. In fact, dry scratch-brushing is very useful in many cases in which it is desirable to impart an artificial coating of brass upon an article to which silver or gold will not readily adhere. In scratch-brushing without the employment of beer, or some other liquid, however, great care must be taken not to continue the operation too long, as the minute particles of metal given off by the scratch-brush would be likely to prove prejudicial to the health of the operator were he to inhale them to any great extent.

62. In gilding or silvering steel articles by the processes described in paragraph 55, it will be necessary to clean them with a little diluted potash in order to remove grease; and if the articles are polished steel, the friction applied must be brisk, when first employing the solution, to ensure an even coating. Penknives, scissors, razors, and other similar goods may be lightly coated with either gold or silver by the solutions described, but it must be understood that the coating is more for beauty than for wear. If properly done, however, the coating will last a considerable time without removal. Needles gilt and silvered in this way are very agreeable to work with, and form an elegant article for ladies' use.

63. As cyanide of potassium, which is so extensively employed in electro-metallurgy, is a highly poisonous substance, the apartments in which electro-gilding and plating are conducted should be well ventilated, and the solutions always be placed in such a position as to be nearest the window or chimney. Breathing an atmosphere impregnated with the vapour arising from the gilding and plating baths is unquestionably prejudicial to health, therefore the reader cannot be too particular in the matter of ventilation. A little liquid ammonia occasionally poured on the floor is advisable, especially in very hot weather.

64. In reducing old solutions by means of sulphuric acid, it must be borne in mind that the fumes are highly poisonous, and must not be inhaled in the *slightest* degree. The operation should be conducted either in the open air, or where there is a good draught to carry off the fumes. Beyond the current created by an open window and door, there should be no motion in the apartment, so that the fumes may be allowed to escape without being diffused. It must be remembered, however, that the fumes (consisting greatly of carbonic acid) are very dense, consequently it will be necessary to allow them to escape from the *lower* part of an apartment; they must not be expected to make their exit from an aperture *above* them.

65. In "colouring" gold of inferior quality, that is to say, below the English standard (18-carat gold), the following mixture may be used with success, and, if carefully employed, even 12-carat gold may be coloured by it :—

Take

Nitrate of potassa (saltpetre) . .	4 ounces
Alum	2 „
Common salt	2 „

Add sufficient warm water to mix the ingredient into a thin paste; place the mixture in a small pipkin or crucible, and allow to boil. The article to be coloured should be suspended by a wire and dipped into the mixture, where it should remain from ten to twenty minutes. The article should then be removed, and well rinsed in hot water, when it must be scratch-brushed, again rinsed, and returned to the colouring salts for a few minutes; it is then to be again rinsed in hot water, scratch-brushed, and finally brushed with soap and hot water, rinsed in hot water, and placed in box-sawdust. The object being merely to remove the alloy, as soon as the article has acquired the proper colour of fine gold, it may be considered sufficiently acted upon by the above mixture. The colouring salts should not be used for gold of a lower standard than 12-carat gold, and even for this quality of gold some care must be taken when the articles are of a very slight make.

66. The process of hardening and tempering small steel implements, such as drills, scrapers, gravers, burnishers, &c., may be useful to persons living away from large towns; therefore I have thought it prudent to give an outline of the method to be pursued. The process of hardening and tempering will apply to any steel tool, of any dimensions, but we will take a small chisel as an example. The blade of the chisel should be removed from the haft or handle, and, the lower end being held by a pair of pliers, the blade is to be placed in a brisk fire until red hot; it is then to be removed and allowed to cool for a few moments, and then dipped into cold water. The next thing to do is to file the blade into the required shape, finishing the filing with a smooth file. When this is done (a vessel of cold water being at hand), the blade is to be again returned to the hottest part of the fire, and allowed to become *white* hot, *i.e.*, as hot as a fierce fire will make it; and as soon as this is done withdraw it quickly with the pliers, and plunge it *instantly* into the vessel of cold water. If this has been done properly, on passing a file over the sharpened surface of the blade it will be found to have no effect upon it; in fact, the blade has become so hard that the file will not make any impression upon it. The upper surface of the blade should now be polished by rubbing it upon a sheet of emery-cloth moistened with oil, and then wiped with a piece of rag. The blade now requires to be *tempered*, *i.e.*, reduced in hardness. This is done by holding the sharpened *edge* of the blade with the pliers, and placing it in the fire until the polished surface of the steel changes colour. The surface nearest the handle will soon assume a blue colour, and an orange tint immediately following it, towards the upper end of the blade. At this point great care is required, or the instrument will be overheated and too soft; therefore, as soon as the orange tint makes its appearance within an inch of the sharp edge of the blade, withdraw it from the fire and examine it quickly; if the extreme edge is of a pale straw-colour, dip the blade *instantly* in cold water; but it must be understood that the straw-colour should be visible at the extreme point, or edge, of the tool. When such is the case, the tempering is complete. The next process is to sharpen it upon a good Turkey stone, and it is ready for use. In hardening and tempering very small tools, such as drills, they may be more safely tempered by dipping them, point upward, in hot sand, by which means they are less likely to become over-heated. Some persons prefer dipping such tools into oil rather than water after removal from the fire; others, again, employ dilute sulphuric acid. It must be borne in mind, however, that in hardening steel the greatest heat is required in the first instance, and in the second, the greatest cold we can obtain. Upon these extremes, properly applied, depends the success of the operation.

67. A battery which has been somewhat commended it may be as well briefly to mention. It consists in using a saturated solution of bicromate of potassa, with a quarter part sulphuric acid, in the carbon cell of a Bunsen's battery (see p. 80); and dilute sulphuric acid, or saturated solution of sal ammoniac, in the zinc cell. Although this battery, or rather modification, has some advantages, it has yet to be further developed before it can safely be recommended for practical purposes.

68. A process of coating metals has been suggested by M. Weil, and for some purposes it may be employed with advantage. M. Weil's object is to avoid the use of cyanide of potassium and the battery, the first being deleterious, and the second expensive. His process may be briefly described as follows:—Instead of employing cyanide, he takes solutions of oxide of metals with the addition of organic matter—tartaric acid, albumen, or glycerine, to prevent the precipitation of the oxides by the fixed alkalies. A solution thus formed may be used either hot or cold. For a coppering solution, he recommends the following formula :—

Sulphate of copper	350 grammes.
Dissolve in hot water and allow to cool . .	10 litres.
Crystallized Rochelle salt (potassio-tartrate of soda)	1,500 grammes.
Caustic soda (containing 50 to 60 per cent. free soda) *	800 „

The articles are to be suspended by zinc wires, or a thin strip of zinc may be attached to them when in solution; and when it is remembered that this process is independent of the battery, the solution will require renewing from time to time. To accomplish this, M. Weil proceeds as follows :—

The zinc which the solution has acquired must be precipitated by sulphide of sodium (prepared by fusing soda with sulphur) gradually, as an excess dissolves the precipitate formed by the sulphide of soda. When the zinc has been precipitated as above directed, the solution must be allowed to settle, and the clear liquor poured off. The solution is then to be resupplied with a solution of sulphate of copper, as before.

It is stated that copper deposited in this way adheres very firmly to iron.

Zincing by the above process is carried out by forming a concentrated solution of potassa or soda, which must be heated to 100° C., a piece of metallic zinc being placed in the solution in contact with the article to be coated.

True bronze, that is, a mixture of tin and copper, may be deposited by the above process, by adding to

* Caustic soda is prepared by dissolving 12 parts of ordinary soda in hot water, and then adding, gradually, 2 parts of lime recently slaked. The liquor should be boiled for an hour, and allowed to cool and settle; the clear liquor being decanted for use.

the copper bath before described stannate of soda or bichloride of tin, previously treated with the solution of soda. The metal to be coated must be in contact with zinc.

69. A new constant battery has been suggested by Mr. A. Reynolds, in which a solution of perchloride of iron is used as an exciting fluid, and metallic iron as the positive electrode.

70. Copper may be coated with antimony, as follows:—Dissolve 1 ounce protochloride of antimony (butter of antimony as it has been called) in 1 pint of spirit of wine; now add hydrochloric acid until the solution is clear. The article to be coated, being previously cleaned, will receive a bright coating in about an hour. Cast iron must be previously coppered by any of the processes before described, and will then receive a coating of antimony from the above solution.

71. As the refining of gold and silver is so closely connected with the electro-plater's art, the following hints may prove serviceable, it is hoped, to those unacquainted with the art of separating the precious metals from their alloys. An alloy of gold, silver, and copper, should be thus treated:—Suppose the alloy to be what is called "jewellers' gold," or the material with which cheap jewellery is made, and which, being unfit for use, is "only fit for the melting-pot." The alloy is first to be melted in a crucible with one-third of its weight of silver, a little dried potash or borax being used as a flux; when the alloy is well melted, it is to be poured into a vessel of cold water, which must be briskly stirred during the operation; and it is well to have a few small pieces of straw, or short sticks, floating on the surface of the water at the time, the object of which is to assist the process of *granulation*. The alloy will now be found at the bottom of the vessel in small grains, which must be carefully collected, and in order to be certain that there is no waste, the alloy should be weighed before and after melting and granulation. With ordinary care scarcely any difference will be observed. The grains must now be put into a clean Florence flask, or other suitable vessel, and dilute nitric acid (1 part acid to 2 parts water) poured upon them: the grains should be allowed to digest for several hours; and, in order to promote chemical action, it will be advisable to place the flask upon a "sand-bath" or near the fire, but this may not be necessary until towards the end of the operation. The gold will now be found at the bottom of the flask, in the form of a brown powder or brown spongy lumps; but, in order to secure the entire removal of the alloys, it will be necessary to decant the solution of silver and copper in the nitric acid into a separate vessel, to be afterwards treated, and fresh nitric acid should then be poured on to the gold, and heat applied as before, to ascertain whether the alloys have been effectually removed. If *red fumes* are still given off in the flask, the separation has not been complete, and the action must be kept up until the red fumes cease to appear even at a boiling temperature. When this is the case, the acid solution must be again poured off, and the gold well washed with hot water, to remove any trace of silver or copper which may be in its interstices, especially if it is in a spongy form. At this stage the brown deposit or mass of gold is pure, and merely requires to be melted into a button with borax or potash.

The solution of silver and copper may next be

treated—the silver being thrown down by strips of copper; and, to ascertain whether *all* the silver has been precipitated by the copper, a small quantity of the solution may be placed in a glass, and a drop or two of hydrochloric acid or a solution of common salt applied; when, if any silver remains in solution, it will assume a milky appearance; if such is not the case, the cupreous liquor may be poured off, and the reduced silver well washed with hot water several times. The silver should then be dried and fused, a little dried potash being mixed with it previous to placing it in the crucible. The silver may then be granulated, as before, or cast in a mould of suitable size; and, being rolled out, will serve as an anode.

It must be understood that the above is but a trifling sketch of the principles of refining in the moist way; yet it is hoped that the reader will learn therefrom enough to enable him to pursue the art of separating gold and silver from their alloys for his own purposes.

When gold and silver are alloyed, or mixed with copper, brass, iron, &c., as in the case of jewellers' waste, filings, &c., the waste, being previously burned in an iron pan to destroy any organic matter present, should be well mixed with a little dried potash, and melted in a crucible. When perfectly fused, a few crystals of nitrate of potash (nitre of commerce) must be dropped into the crucible from time to time, which will remove from the gold and silver whatever copper or iron may be present, if sufficient nitre has been employed. The nitre must, however, be added cautiously; otherwise, if there be organic matter present, the flux may rise above the melting-pot, and, overflowing, carry part of the metal with it. It will be necessary to watch the operation closely, and if the flux, &c., appear likely to overflow, a small quantity of dried common salt thrown into the crucible will check the ebullition, and tend to keep the metals and flux to the lower part of the vessel. When the operation is complete, the melting-pot or crucible must be withdrawn from the fire and set aside to cool. When cold, the pot must be broken at the base with a hammer, and the "button" of metal withdrawn. The button must now be again melted, as before described, with borax or potash, and granulated, the grains being treated as before, to remove the silver and copper.

As the above details are intended for the use of those who may be unacquainted with the art of refining, it is hoped that they may be found sufficient to enable the beginner to commence a study of a most interesting and important branch of industry.

72. Cyanide should seldom be added to a bath, whether of gold or silver, when the anode, being at work, is clean and uniform in appearance. It is always objectionable to have too great an excess of this salt in the bath; therefore it should never be added until the batteries and their connections have been well examined. Sometimes, when the battery is somewhat exhausted, the anodes become slightly discoloured, especially if a larger surface of goods is exposed than is proportionate to the surface of anode; in such case, it will be necessary to increase the activity of the battery rather than to add cyanide to the bath. Cyanide is a good friend, but a bad foe. All that a good bath requires is a slight excess of cyanide, if the battery is in good order and but little organic matter has accumulated in the bath. When, on the other hand, the bath has acquired a good deal of organic matter (under which condition it is generally preferable to a new bath), it will require a greater excess of cyanide, and this excess, under such circumstances, will be very beneficial. An old or well-worked bath will bear a much larger amount of cyanide, in proportion, than a new one.

73. Old steel dessert-knives, which have been "close-plated," or solder-plated, as it is termed, will sometimes give the operator a good deal of trouble before he can deposit a sound coating of silver or gold upon them. Suppose the old silver has been stripped off, and the solder removed, the author has more than once found that a very serious difficulty has arisen when the articles have been coated either with gold or silver, or with copper, previous to coating with the other metals. For instance, a dozen dessert-knives, having been stripped and well cleaned, were placed in the alkaline copper bath. After an hour's immersion, the articles were removed from the bath and examined, when it was found that in every part of each blade, from heel to point, the blades were found to be *cracked,* exhibiting fissures in some cases nearly $\frac{1}{16}$th of an inch wide. These cracks pervaded each blade in almost every part, and it was some time before the cause of these remarkable flaws could be traced. In examining the interior of the cracks, it was found that the copper was freely deposited, and, the coating becoming thicker and thicker, the deposit of copper *forced open* the cracks (although at first invisible), until they assumed the alarming appearance we have described. It must be remembered that the numerous *points* which the fractured metal presented would greatly favour the deposit in those parts, in pre ference to the plain surfaces, and hence the opening of the invisible flaws in the metal as the deposit thickened. The same result has occurred in certain articles of wrought steel of bad quality; and when we bear in mind that the processes of grinding and polishing disguise such defects in steel, it must not be expected that they will show themselves until some time after deposition has begun. In order, however, to prevent, as far as possible, the flaws we mention from opening during deposition, we recommend that the first coating should be allowed to take place almost immediately after the article is immersed in the bath, and as soon as this is done, the article should be quickly rinsed, scratch-brushed, and returned to the bath. Again, it will be advisable to employ a tolerably weak bath at first, with good battery power, and after the articles have received a preliminary coating under these conditions, they may then be more safely allowed to remain in the bath until the required amount of metal is deposited.

74. When wooden vessels are employed for plating, they should be well saturated in boiling water for a few hours, or even days, before the solution of silver is poured in; as, independent of the absorption of silver by the wood, if it is well moistened with water, the interstices are not so readily acted upon by cyanide.

75. Gutta-percha vessels or linings of gutta-percha should never be used for a silver bath under any circumstances, as the cyanide acts upon gutta-percha and the adulterants with which it is always, more or less, contaminated. Nitrate of silver, also, acts upon gutta-percha, as most photographers are aware; and, therefore, it will be absolutely necessary to avoid using gutta-percha in contact with either cyanide of potassium or nitrate of silver. I have known more than one instance in which a silver bath, being destroyed by a gutta-percha lining, has been reduced by sulphuric acid; and, soon after the acid has been applied, the gutta-percha has been set free, and has floated in the solution in clots of considerable size.

76. When the operator's hands have become injured by coming in contact with cyanide (which is frequently the case if there is an abrasion of the skin), it is a good plan to dip them for a few moments in very dilute sulphuric acid (20 drops in a tumbler of water), and then rinse them well. The hands should then be well soaked in tolerably hot water, well dried, and finally saturated with oil or grease of some kind. The author has frequently suffered from sores occurring under each nail of both hands, in consequence of neglecting to wash the hands immediately after immersion in a solution of cyanide. The electro-plater should be careful to avoid this, as sores thus formed are difficult to heal, and cause great pain to the part affected.

77. Copper or brass wire should never be bent or twisted more than once or twice without being annealed. When the wires used for slinging goods, or the wires proceeding from the electrodes of a battery, have been used more than once, they are apt to become brittle where they have been bent, and are liable to break; it is better, therefore, occasionally, to make the wires red hot, and, when cool, stretch them out by drawing them several times across the edge of a board, by which means they will readily become straightened. The wires should then be passed through a piece of emery-cloth, to clean them. The ends of slinging wires, however, which have become coated with silver, should first be dipped for a few moments in the hot stripping solution (p. 56), and finally treated as above.

78. The moment the zinc plate of a battery evolves gas, accompanied by a hissing noise, the plate should be withdrawn and at once re-amalgamated; for, as soon as local action sets in, the current becomes greatly diminished; added to which, if allowed to continue, the zinc plate will soon be destroyed, without accomplishing its task. It is not uncommonly the case that local action begins on the first day that a battery is set in action; and, therefore, in order to remedy this defect as early as possible, the battery should be carefully watched, and the zinc plate removed the instant effervescence is observed in the battery-cell.

79. In order to ascertain whether an article is made of gold, if a doubt arises, a simple plan is to rub a portion of the article upon a piece of slate, Wedgewood ware or Turkey stone, and then apply a single drop of nitric acid by touching the part with the stopper of the bottle. If the acid produces no effect, the article may be considered gold. A very inferior alloy of gold, however (12-carat gold), will stand this test; but its colour will act as a guide, as it will fail (except when electro-gilt) to present the rich yellow colour of good gold. When a "common gold" article has been strongly gilt, it will be advisable to pass a keen but smooth file over a small part of the article, and then apply the nitric acid to the part, when, if the article is not genuine, the characteristic green tint of nitrate of copper will at once show itself. As it is commonly the practice to designate articles manufactured from "plated" metal (*i.e.*, gold and metal united and rolled out into thin sheets) "fine gold," the electro-gilder should make himself acquainted, if he is not so already, with the various kinds of genuine and spurious gold in commerce ; otherwise, should an accident occur to an article, he may lose more than he has a right to do; and, again, *plated* articles require a somewhat different treatment to that which is applied to articles of genuine gold. In applying the term "genuine gold," we suppose we must be understood to mean such articles as are made of an alloy of gold which *will* stand the test of nitric acid, if only to contradistinguish them from plated-gold articles. An instance once occurred to the author in which, by mistake, he placed two "gold" brequet or Albert chains in a jar containing nitric acid, instead of dipping them into a vessel of warm water, which stood beside it. Being away for an hour, on returning to the gilding-room it was found that the nitric acid had entirely removed all trace of the brequets, except a thin shell of one of the links, which floated upon the surface of the acid, and a dark-brown powder which had deposited in the vessel. On pouring off the acid, and washing the precipitate, which was found to be almost pure gold, it at once became evident that the so-called gold brequets (which had been invoiced at 55*s*. 6*d*. each, wholesale price) were, in fact, made of *plated* metal, but of so good a quality that a very good judge might readily have been deceived. The object of the above remarks is to place the electro-gilder on his guard. He should also endeavour to satisfy himself whether an article is really gold, or an alloy of gold, or plated, before preparing it for the bath, as, if it is gold, it will require as a rule a rather stronger current, and a larger surface of anode exposed in solution, than an article made of plated metal.

80. The back parts or hollows of casts, or "struck" work, will sometimes be troublesome to gild or silver, more especially because these surfaces must of necessity be kept at a distance from the anode, and besides which a concave surface (even if placed directly facing the anode) always receives the deposit more tardily than a convex surface. It will be necessary then, in order to aid the deposition upon the hollow surfaces, to keep the articles moved in the bath when first immersed, by which means all surfaces will receive the deposit alike, and the deposition which takes place after will be in the order required—that is, the outer surfaces will receive the greatest amount of metal, and the inner surfaces, or hollows, the least, but still sufficient for all purposes. If this precaution is not adopted, it is quite possible that the exposed surface of the article will be well coated, whilst the hollows will scarcely be coated at all. Nothing ensures uniformity of deposit so much as gentle motion in the bath, and in some manufactories an apparatus has been applied for keeping articles, while suspended in solution, in a constant but gentle state of motion—a practice very highly to be commended, upon principle, if it can be carried out with economy.

81. The effect of *motion* whilst an article is receiving the deposit is most clearly seen during the operation of gilding. If a watch-dial, for instance, is placed in the gilding-bath, and allowed to remain for a few moments undisturbed, if the solution of gold has been much worked, it is probable that the dial will acquire a dark red or "foxy" colour; but if it is quickly moved about, it instantly changes colour, and will sometimes even assume a pale-straw colour. In fact, as we have before observed, the colour of a deposit may be regulated greatly by motion of the article in the bath—a fact which the operator should study with much attention when gilding. In depositing brass from its solutions, the effects of motion are even more remarkable; for, by keeping the articles moved, copper alone will be deposited, whilst, on the other hand, if stationary, a proper alloy of copper and zinc will be obtained.

82. As there is always a deposit or sediment of some nature at the bottom of a bath, whether of gold or silver, the articles to be coated should never be immersed so deeply in the solution as to come in contact with this deposit. If spoons and forks, for instance, are allowed to reach nearly to the bottom of the bath or solution-vat, every time they are lifted up or lowered the small particles which had settled at the bottom of the vessel become disturbed, and resting upon the articles, will prevent the deposition of silver taking place wherever these particles are present, thus causing an irregularity of surface. As it frequently happens that small particles of silver fall off the surface of the anode, these will sometimes, if the sediment is disturbed, as we have pointed out, rest upon the lower parts of the work, and the deposit will take place *over* them, and when submitted to the scratch-brush the silver will strip or peel off; or, if the deposit is very thick, the bowls of spoons or prongs of forks, or such surfaces as are nearest the bottom of the bath, will be exceedingly rough, more especially as deposition always takes place more freely at the lower surface of the solution. There should, therefore, always be a certain distance between the lower end of articles in solution and the bottom of the vessel in which they are coated.

83. As pure silver is more easily oxidised, or tarnished, than standard silver, electro-plated articles should always be carefully protected from the atmosphere, especially if it be moist or vitiated. Electro-plated goods must always be kept well wrapped up, and in a perfectly dry situation; otherwise oxidation soon sets in, and the goods become unsightly and unsaleable.

84. Although a silver bath is improved by acquiring a moderate amount of organic matter, yet the operator should be careful not to suffer this accession to the bath to take place either too suddenly or in large quantities at a time. For instance, if candlesticks, which generally are filled with a compound of rosin or pitch, are placed in the solution without being previously freed from these substances, the cyanide will dissolve a considerable portion of the composition, and the conductibility of the solution is thereby lessened to some extent. In a moderate degree, the presence of such matter in a bath is an advantage; but it should never be allowed to enter the bath in large quantities at a time. A new bath which suddenly acquires a quantity of organic matter in the way we have described is apt to work sluggishly and with irregularity, and not unfrequently the deposit becomes coarse and spotted. On the other hand, a bath which in the course of several months has absorbed a small amount of organic matter (which will cause it to assume a dark-reddish colour) will give a much finer and brighter deposit than a bath newly made; and, at the same time, goods plated in it will be less liable to strip than when plated in a new bath.

The same observations do not apply to a gilding bath, which is generally worked hot, and, besides which, the colour of the deposit is of the first importance. The presence of organic matter in a gold bath tends to deepen the colour; and if in great excess, the work will frequently assume a foxy-red tint, and this, in some cases, would be highly prejudicial. It is better, therefore, to keep the gold solution as free as possible from organic substances; and as one of the chief causes of a gold bath acquiring organic matter is the *imperfect rinsing of articles after scratch-brushing*, whereby the beer used in the operation of scratch-brushing becomes washed out of the interstices of the article when in the bath, it is advisable not only to rinse all articles well before gilding, but in fact they should always receive a final rinsing in perfectly clean hot water. It is not an uncommon practice for electro-gilders to rinse many articles, after being scratch-brushed, in the same water, and to transfer them directly to the gilding bath. Now this is highly objectionable: the water should be frequently changed, and when we consider the difference between investing a little trouble in renewing the rinsing water and making up a fresh gold solution, it will at once become apparent that the former will be most likely to yield an advantage. Again, a gold solution which has been worked a long time without becoming discoloured (the discoloration generally being due to organic matter) is far better than a new solution, and, for most purposes, will produce better results; but as soon as the solution becomes charged with the impurities referred to, its action is uncertain and irregular. It is to be hoped that the less practical reader will bear these observations in mind, that he may experience as little disappointment as possible in his operations.

85. As lead edges or mounts of cruet-frames, candlesticks, soy-frames, and similar plated goods are very troublesome to electro-plate, except in the hands of a very experienced person, it is frequently advantageous to adopt a plan commonly pursued, namely, to have the edges cast in brass or German silver, the old edges or mounts removed, and the newly-cast edges soldered upon the article. By this means all difficulty is removed, and the article, when finished, is not only rendered more durable, but can also be more highly finished.

86. It is a good plan for the operator to secure impressions in gutta-percha of all mounts, or parts of the same, which are likely to become serviceable to him at an after-time. These impressions can be electrotyped by the processes described in the early part of this work, and at any time castings can be made from them which may be applicable to many useful purposes. For instance, if an impression of a few inches of a gadroon mount be taken, and an electrotype obtained therefrom, and a few castings made from the electrotype, whenever the operator requires a few inches or even a part of an inch of such a mount, to supply the place of a broken gadroon edge, he will find the castings of great service to him. And by adopting this practice, it is possible not only to accumulate copies of very useful, but of many very choice mounts. And from these electrotype copies many beautiful and useful articles may be formed by carefully grouping the impressions obtained from various articles, or parts of them, to form a new design of an entirely novel character. The author has produced some very pleasing effects by arranging alternately, for instance, a piece of scroll-work an inch and a quarter in width, and a cornucopia of equal width, the latter being about two inches long. Four impressions of each being taken, and soldered together, formed a handsome octagonal salt-cellar. A circular ring of wire, connected by short pieces of wire to the inner surface of each "upright," formed a resting-place for the glass or salt vessel, which may be made either of blue, red, or white glass, according to taste. From the above hints the reader may glean sufficient to enable him to construct, upon the same principle, many articles of great beauty, and at the same time possessing novelty of construction, if not of design. Impressions of very rare subjects may be taken and applied in this way, and there is scarcely a limit to the variety of effect which may be produced.

87. A very pleasing effect may be produced upon a plain silver article, by sketching a design upon it with a good lead pencil—a name or initials, for example—and if the article is then placed in the gilding-bath for a few moments, all parts which have not been traced by the lead pencil, will have become gilt. When the article has been rinsed, gentle rubbing with the finger will remove the plumbago, or "black lead," and beneath the design will appear in silver. By reversing the operation, the design may be made to appear in gold. Plumbago answers well for this purpose, owing to its being an inferior conductor to either gold or silver, especially in alkaline solutions. In practising this process, however, care must be taken not to move the articles while in solution, otherwise the plumbago may be worked off by friction, and, consequently, the design will be obliterated by deposition taking place where the design appeared.

88. When several articles made of different metals

are to be plated in the same bath, the article which is the worst conductor should be put in first. Thus, supposing copper, brass, and German-silver articles are to be immersed, the copper article should be suspended first, the brass next, and the German silver last. At the same time, the anode must be slightly lowered as each article is suspended. If convenient, it is better to plate articles composed of one metal or alloy only, at the same time.

89. Bichloride of mercury in solution has been recommended for amalgamating zinc plates instead of the ordinary method, and, in many respects, it would appear to present an advantage. But if zinc plates are cast from a mixture of zinc and mercury, they would be still more effective and durable, and less liable to local action. The alloy of zinc and mercury consists in melting zinc in the ordinary way, occasionally adding a little grease, rosin, or sal ammoniac, and, when melted, pour in gradually mercury in the proportion of—1 ounce of mercury to each pound of zinc. Zinc thus alloyed is exceedingly brittle, and the plates will therefore require careful treatment. If well prepared, however, these plates possess many advantages over the ordinary amalgamated plates.

90. All old plated articles which have to be pumiced or rendered smooth with Water-of-Ayr stone, or otherwise prepared for plating with the aid of water (that is to say, not being rendered smooth by emery-cloth, &c.), should have a vessel kept specially for them; for instance, a wooden tub with a board placed across, in order that the particles of silver rubbed off by the pumice, &c., may be collected in the vessel beneath; and the electro-plater cannot be too careful in saving this and all other kinds of waste, either of gold or silver. Many persons who have omitted to save this kind of waste from the commencement of their operations have nurtured the idea that it was useless (having lost so much already!) to begin to save such comparatively trifling waste; but we would strongly impress upon the student the great importance (when dealing with the precious metals) of using the utmost economy. Gold and silver are always recoverable, in some shape or other; therefore, to allow them to escape down a gutter beyond the power of recovery (as many photographers do) is not only foolish, but wicked.

91 In commencing the study or practice of electro-deposition, after reading this work, the first question that will occur to the student is, "What apparatus and chemicals shall I require?" Therefore, well knowing how useful such information in a brief form will be, we have thought it advisable to give the subjoined list of things which are either indispensable or may become necessary, either to the student who works for pleasure or to those who work for profit.

92. LIST OF ARTICLES REQUIRED IN ELECTRO-GILDING, PLATING, ETC.

Gilding battery-jar.
Plating battery-jar.
Gilding bath, of glass or stone-ware.
Plating bath, of wood or stone-ware.
Gold anode rolled out to a moderate thickness.
Silver anode rolled out to a moderate thickness.
1 lb. of stout copper wire.
1 lb. of thin copper wire for slinging.
One or two Bath bricks, to be rubbed together until powdered, or
A few pounds of powdered pumice-stone.

Several brushes, consisting of 1, 2, 3, and 4 rows each.
Several sheets of emery-cloth, Nos. 1, 2, and 3.
Several lumps of pumice-stone.
A Water-of-Ayr stone about three-quarters of an inch square.
Pair of flat pliers.
Several files.
Chamois leather.
1 ounce of rouge.
½ lb. of mercury.
Several binding-screws.
A few sheets of filtering-paper
1 quart of box-sawdust.
Several scratch-brushes, which, for economy sake, may be cut in half, and the ends soldered.
Scratch-brush lathe and "chuck."
Evaporating dish to hold half a pint.
Rotten stone, 1 lb.
Borax, 1 ounce.
Silver solder.
Soft solder.
Blowpipe.
Rosin.
Soldering iron.
Cyanide of potassium for silver bath.
Cyanide of potassium for gilding bath.
Nitric acid.
Hydrochloric acid.
Sulphuric acid.
Sulphate of copper for electrotyping.
Burnishers.
Nitrate of mercury (made by dissolving mercury in nitric acid).
One or two "buffs" for polishing.
Charcoal, several pieces.
One or two glass measures.
Nitrate of potash for stripping solution.
Fuming nitric acid.
Acetic acid.
Sal ammoniac.
Scales and weights—small and large.
Sheet copper for gilding battery.
Sheet copper for plating battery.
Stout sheet zinc for plating battery.

Stout sheet zinc or cast-zinc bar for gilding.
Ox gullet, or porous cell.
Carbonate of potassa.
Fine silver for solutions.
Fine gold for solutions.
A few gallons of distilled or rain water.
Bisulphide of carbon for "bright" plating.
Common salt.
Caustic soda (see p. 48).
Silver sand.
Jar for stripping solution (see p. 56).
Plumbago, and a camel's-hair pencil.
A tub, or other vessel for cleaning work.
Several pans or rinsing vessels.
Brass rods to suspend articles to be plated.

FRENCH MEASURES OF WEIGHT.

		English grains.
Milligramme	=	·0154.
Centigramme	=	·1543.
Decigramme	=	1·5434.
Gramme	=	15·4340.

MEASURES OF VOLUME.

1 litre = about 34 English fluid ounces.

INDEX.

THE END.

WEALE'S SERIES

OF

RUDIMENTARY, SCIENTIFIC, EDUCATIONAL,

AND

CLASSICAL WORKS.

The entire Series is freely illustrated with wood and steel engravings and lithographs where requisite.

PUBLISHED BY

VIRTUE & YORSTON,

12 DEY STREET,

NEW YORK.

FOR SALE BY ALL BOOKSELLERS.

RUDIMENTARY SERIES.

2.	**Natural Philosophy,** by Charles Tomlinson	$0 40
12.	**Pneumatics,** " "	60
20.	**Perspective,** by George Pyne	80
27.	**Painting; or, A Grammar of Coloring,** by G. Field	80
40.	**Glass Staining,** by Dr M. A. Gessert, with an Appendix on the Art of Enamelling	40
41.	**Painting on Glass,** from the German of Fromberg...	40
50.	**Law of Contracts for Works and Services,** by David Gibbons............................	60
66.	**Clay Lands and Loamy Soils,** by J. Donaldson....	40
69.	**Music,** Treatise on, by C. C. Spencer	80
71.	**Piano-forte, Art of Playing,** by C. C. Spencer	40

72. **Recent and Fossil Shells** (a Manual of the Mollusca), by S. P. Woodward, F.G.S., A.L.S., etc. New edition, with Appendix by Ralph Tate, F.G.S. Cloth extra, gilt top........................ $3 00
79**. **Photography,** Popular Treatise on, from the French of Monckhoven, by W. H. Thornthwaite.. 60
96. **Astronomy**, by the Rev. R. Main. New and Enlarged edition; with an appendix on "Spectrum Analysis"................................ 60
108. **Metropolis Local Management Acts** (London).. 60
108*. " " " **Amendment Act, 1862.** With Notes and Index........... 40
Nos. 108 and 108* in One Vol 1 00
112. **Domestic Medicine,** by Dr. Ralph Gooding........ 80
112* **Management of Health.** A Manual of Home and Personal Hygiene. By James Baird, B.A........... 40
113. **Use of Field Artillery on Service,** by Jaubert, translated by Lieut-Col. H. H. Maxwell 60
113*. **Memoir on Swords,** by Marey, translated by Lieut.-Col. H. H. Maxwell............................ 40
150. **Treatise on Logic,** by S. H. Emmens, Esq.......... 60
151. **Handy Book on the Law of Friendly, Industrial and Provident, Building and Loan Societies.** With Copious Notes. By Nathaniel White. 40
152. **Practical Hints for Investing Money:** with an Explanation of the Mode of Transacting Business on the Stock Exchange. By Francis Playford......... 40
153. **Locke's Essay on the Human Understanding,** Selections from, by S. H. Emmons.............. 80
163. **The Law of Patents for Inventions,** by F. W. Campin, Barrister............ 80

PHYSICAL SCIENCE.

1. **Chemistry,** by Prof. Fownes, including Agricultural Chemistry, for the use of Farmers.................. $0 40
3. **Geology,** by Major-Gen. Portlock.................. 60
4. **Mineralogy,** with a Treatise on Mineral Rocks or Aggregates, by Dana.......................... 80
7. **Electricity,** by Sir W. S. Harris.................. 60
7*. **Galvanism, Animal and Voltaic Electricity,** by Sir W. S. Harris.............................. 60
8. **Magnetism,** Exposition of, by Sir W. S. Harris..... 1 40
11. **Electric Telegraph,** History of, by R. Sabine, C. E. 1 20
133. **Metallurgy of Copper,** by R. H. Lamborn......... 80
134. **Metallurgy of Silver and Lead,** by R. H. Lamborn...................................... 80
135 **Electro-Metallurgy,** by A. Watt 80
138. **Hand-book of the Telegraph,** by R. Bond....... 40
143. **Experimental Essays**—On the Motion of Camphor and Modern Theory of Dew, by C. Tomlinson...... 40
161. **Questions on Magnetism, Electricity, and Practical Telegraphy,** by W. McGregor. 60

BUILDING AND ARCHITECTURE.

16. **Architecture,** Orders of, by W. H. Leeds 40
17. " Styles of, by T. Bury 60
18. " Principles of Design, by E. L. Garbett. 80
22. **Building,** the Art of, by E. Dobson................ 60
23. **Brick and Tile Making,** by Dobson and Mallett... 1 20
25. **Masonry and Stone-Cutting,** by E Dobson 1 00
30. **Draining and Sewage of Towns and Buildings,** by G. D. Dempsey 80
With No. 29, Drainage of Land, 2 vols. in 1...... 1 20
85. **Blasting and Quarrying of Stone, and Blowing up of Bridges,** by Lt.-Gen. Sir J. Burgoyne....... 60

MACHINERY AND ENGINEERING.

47. **Lighthouses,** their Construction and Illumination, by Allan Stevenson ... $1 20
59. **Steam Boilers,** their Construction and Management, by R. Armstrong ... 60
62. **Railways,** Construction, by Sir M. Stephenson ... 60
62*. **Railway Capital and Dividends,** with Statistics of Working, by E. D. Chattaway ... 40
(Vols. 62 and 62* bound in one) ... 1 00
67. **Clocks, Watches and Bells,** by E. B. Denison ... 1 40
Appendix (sold separately) ... 40
78. **Steam and Locomotion,** on the Principle of connecting Science with Practice, by J. Sewell ... 80
78*. **Locomotive Engines,** by G. D. Dempsey ... 60
78*. **Illustrations to the above.** 4to ... 1 80
98. **Mechanism and Construction of Machines,** by T. Baker; and **Tools and Machines,** by J. Nasmyth, with 220 Woodcuts ... 1 00
114. **Machinery,** Construction and Working, by C. D. Abel ... 60
115. **Plates to the above.** 4to ... 3 00
139. **Steam Engine,** Mathematical Theory of, by T. Baker. 40
155. **Engineer's Guide to the Royal and Mercantile Navies,** by a Practical Engineer ... 1 20
162. **Brassfounder's Manual,** by W. Graham ... 1 00
164. **Modern Workshop Practice,** as applied to Marine, Land, and Locomotive Engines, Floating Docks, Dredging Machines, Bridges, Cranes, etc., by J. G. Winton ... 1 20

CIVIL ENGINEERING, ETC.

13. **Civil Engineering,** by H. Law and G. R. Burnell ... 1 80
29. **Draining Districts and Lands,** by G. D. Dempsey .. 60
(With No. 30, Drainage and Sewage of Towns, 2 vols. in one) ... 1 20
31. **Well-Sinking, Boring, and Pump Work,** by J. G. Swindell, revised by G. R. Burnell ... 40
46. **Road-Making and Maintenance of Macadamized Roads,** by Gen. Sir J. Burgoyne ... 60
60. **Land and Engineering Surveying,** by T. Baker ... 80

63. **Agricultural Engineering, Buildings, Motive Powers, Field Engines, Machinery, and Implements,** by G. H. Andrews $1 20

77*. **Economy of Fuel,** by T. S. Prideaux 60

80*. **Embanking Lands from the Sea,** by J. Wiggins.. 80

82. **Water Power,** as applied to Mills, etc., by J. Glynn.. 80

82**. **Gas Works and Manufacturing Coal Gas,** by S. Hughes 1 20

82***. **Water-Works for Cities and Towns,** by S. Hughes 1 20

117. **Subterraneous Surveying, and Ranging the Line without the Magnet,** by T. Fenwick, with Additions by T. Baker 1 00

118. **Civil Engineering of North America,** by D. Stevenson 1 20

120. **Hydraulic Engineering,** by G. R. Burnell 1 20

121. **Rivers and Torrents,** and a Treatise on **Navigable Canals and Rivers that Carry Sand and Mud,** from the Italian of Paul Frisi 1 00

125. **Combustion of Coal, and the Prevention of Smoke,** by C. Wye Williams, M.I.C.E. 1 20

SHIP-BUILDING AND NAVIGATION.

51. **Naval Architecture,** by J. Peake 1 20

53*. **Ships for Ocean and River Service,** Construction of, by Captain H. A. Sommerfeldt 40

53**. **Atlas of 15 Plates to the above,** Drawn for Practice. 4to 3 00

54. **Masting, Mast-Making, and Rigging of Ships,** by R. Kipping 60

54*. **Iron Ship-Building,** by J. Grantham. Fifth edition with supplement $1 60

54**. **Atlas of 40 Plates** to the preceding. folio...... 15 20

55. **Navigation;** the Sailor's Sea Book: How to Keep the Log and Work it off, etc.; Law of Storms, and Explanation of Terms, by J. Greenwood............ 80

80. **Marine Engines, and Steam Vessels, and the Screw,** by R. Murray........................... 1 00

83 *bis.* **Ships and Boats,** Forms of, by W. Bland......... 60

99. **Nautical Astronomy and Navigation,** by J. R. Young.. 80

100*. **Navigation Tables,** for Use with the above....... 60

106. **Ships' Anchors for all Services,** by G. Cotsell..... 60

149 **Sails and Sail-Making,** by R. Kipping, N.A....... 1 00

ARITHMETIC AND MATHEMATICS.

6. **Mechanics,** by Charles Tomlinson.................. 60

82. **Mathematical Instruments, their Construction, Use, etc.,** by J. F. Heather...................... 60

61*. **Ready Reckoner** for the Measurement of Land, Tables of Work at from 2*s*. 6*d*. to £1 per acre, and valuation of Land from £1 to £1,000 per acre, by Arman 60

76 **Geometry, Descriptive,** with a Theory of Shadows and Perspective, and a Description of the Principles and Practice of Isometrical Projection, by J. F. Heather... 80

83. **Book-Keeping and Commercial Phraseology,** by James Haddon 40

84. **Arithmetic,** with numerous Examples, by J. R. Young. 60

84*. **Key to the above,** by J. R. Young................ 60

85. **Equational Arithmetic,** by W. Hipsley.......... $0 40

85*. **Supplement to Equational Arithmetic,** by W. Hipsley.................................... 40

(Vols. 85 and 85* bound together)............ 80

86. **Algebra,** by J. Haddon............................ 80

86*. **Key and Companion to the above,** by J. R. Young. 60

88. **Euclid's Geometry,** with Essays on Logic, by H. Law. 80

90. **Geometry, Analytical and Conic Sections,** by J Hann.... 40

91. **Plane Trigonometry,** by J. Hann.................. 40

92. **Spherical Trigonometry,** by J. Hann 40

(The two volumes in one)........................... 80

93. **Mensuration,** by T. Baker.......................... 60

94. **Logarithms,** Tables of ; with Tables of Natural Sines, Cosines, and Tangents, by H. Law................ 1 00

97. **Statics and Dynamics,** by T. Baker................ 40

101. **Differential Calculus,** by W. S. B. Woolhouse..... 40

101*. **Weights and Measures of all Nations;** Weights of Coins, and Divisions of Time ; with the Principles which determine the Rate of Exchange, by W. S. B. Woolhouse.. 60

102. **Integral Calculus,** by H. Cox..................... 40

103. **Integral Calculus,** Examples of, by J. Hann........ 40

104. **Differential Calculus,** Examples of, with Solutions, by J. Haddon 40

105 **Algebra, Geometry, and Trigonometry,** First Mnemonical Lessons in, by the Rev. T. P. Kirkman. 60

131. **Ready-Reckoner for Millers, Farmers, and Merchants,** showing the Value of any Quantity of Corn. with the Approximate Value of Mill-stones and Mill Work 40

136. **Rudimentary Arithmetic,** by J. Haddon, edited by A. Arman 60

137. **Key to the above,** by A. Arman.................. 60

147. **Stepping-Stone to Arithmetic,** by Abraham Arman, Schoolmaster, Thurleigh, Beds................... 40

148. **Key to the above,** by A. Arman.................. 40

NEW SERIES OF EDUCATIONAL WORKS.

The standard character of these works, combined with their cheapness, is gradually making them the most popular text-books in the market.

(Single copies of any of our *Educational Works* sent to Professors, School Authorities, and Teachers, at one-half the catalogue price, postage paid, when ordered for examination with a view to their introduction.)

Histories, Grammars, and Dictionaries.	Limp.	Cloth Boards.
	$ c.	$ c.
1. ENGLAND, History of, by W. D. Hamilton	1 60	2 00
5. GREECE, History of, by W. D. Hamilton and E. Levien	1 00	1 40
7. ROME. History of, by E. Levien	1 00	1 40
9. CHRONOLOGY OF CIVIL AND ECclesiastical History, Literature, Art, and Civilization, from the earliest period to the present time	1 00	1 40
11. ENGLISH GRAMMAR, by Hyde Clarke	40	
11*. HAND-BOOK OF COMPARATIVE Philology, by Hyde Clarke	40	
12. ENGLISH DICTIONARY, above 100,000 words, or 50,000 more than in any existing work. By Hyde Clarke		1 80
ENGLISH DICTIONARY, with Grammar		2 20
14. GREEK GRAMMAR, by H. C. Hamilton	40	
15. " DICTIONARY, by H. R. Hamilton. Vol. 1. Greek—English	80	
17. GREEK DICTIONARY, Vol. 2. English—Greek	80	
GREEK DICTIONARY, complete in 1 vol.	1 60	2 00
GREEK " complete in 1 vol., with Grammar		2 40
19. LATIN GRAMMAR, by T. Goodwin	40	
20. " DICTIONARY, by T. Goodwin Vol. 1. Latin—English	80	
22. LATIN DICTIONARY, Vol. 2. English—Latin	60	
LATIN " complete in 1 vol.	1 40	1 80
" " with Grammar		2 20
24. FRENCH GRAMMAR, by G. L. Strauss	40	
25. " DICTIONARY, by A. Elwes. Vol. 1. French—English	40	
26. FRENCH DICTIONARY. Vol. 2. English—French	60	

	Histories, Grammars, and Dictionaries.	Limp.	Cloth Boards.
		$ c.	$ c.
	FRENCH DICTIONARY, complete in 1 vol	1 00	1 40
	FRENCH " with Grammar		1 80
27.	ITALIAN GRAMMAR, by A. Elwes	40	
28.	" TRIGLOT DICTIONARY, by A. Elwes. Vol. 1. Italian—English—French	80	
30.	ITALIAN TRIGLOT DICTIONARY, Vol. 2. English—Italian—French	80	
32.	ITALIAN TRIGLOT DICTIONARY, Vol. 3 French—Italian—English	80	
	ITALIAN TRIGLOT DICTIONARY, complete in 1 vol.		3 00
	ITALIAN TRIGLOT DICTIONARY, with Grammar		3 40
34.	SPANISH GRAMMAR, by A. Elwes	40	
35.	" ENGLISH AND ENGLISH-SPANISH DICTIONARY, by A. Elwes	1 60	2 00
	" " with Grammar		2 40
39.	GERMAN GRAMMAR, by G. L. Strauss	40	
40.	" READER, from best Authors	40	
41.	" TRIGLOT DICTIONARY, by N. E. Hamilton. Vol. 1. English—German—French	40	
42.	GERMAN TRIGLOT DICTIONARY, Vol. 2. German—English—French	40	
43.	GERMAN TRIGLOT DICTIONARY, Vol. 3. French—English—German	40	
	GERMAN TRIGLOT DICTIONARY, complete in 1 vol.	1 20	1 60
	GERMAN TRIGLOT DICTIONARY, with Grammar		2 00
44.	HEBREW DICTIONARY, by Dr. Breslau. Vol. 1. Hebrew—English	2 40	
	HEBREW DICTIONARY, with Grammar	2 80	
46.	HEBREW " Vol. 2. English—Hebrew	1 20	
	HEBREW " complete, with Grammar, in 2 vols.		4 80
46*.	HEBREW GRAMMAR, by Dr. Breslau	40	
47.	FRENCH AND ENGLISH PHRASE BOOK	40	
48.	COMPOSITION AND PUNCTUATION.	40	
49.	DERIVATIVE SPELLING BOOK	60	

GREEK AND LATIN CLASSICS,

With Explanatory Notes in English, principally selected from the best German Commentators.

LATIN SERIES.

1. **Latin Delectus,** with Vocabularies and Notes, by H. Young.......... $0 40
2. **Cæsar's Gallic War**; Notes by H. Young.......... 80
3. **Cornelius Nepos**; Notes by H. Young.......... 40
4. **Virgil.** The Georgics, Bucolics; Notes by W. Rushton and H. Young.......... 60
5. **Virgil's Aeneid**; Notes by H. Young.......... 80
6. **Horace.** Odes and Epodes; Notes, Analysis, and Explanation of Meters.......... 40
7. **Horace.** Satires and Epistles; Notes by W. B. Smith. 60
8. **Sallust.** Catiline, Jugurtha; Notes by W. M. Donne. 60
9. **Terence.** Andria and Heautontimorumenos; Notes by J. Davies.......... 60
10. **Terence.** Adelphi, Hecyra, and Phormio; Notes by J. Davies.......... 80
11. **Terence.** Eunuchus. By J. Davies.......... 60
14. **Cicero.** De Amicitia, de Senectute, and Brutus; Notes by W B. Smith.......... 80
16. **Livy.** Part I. Books i., ii., by H. Young.......... 60
16*. ——— Part II. Books iii., iv., v., by H. Young..... 60
17. ——— Part III. Books xxi., xxii., by W. B. Smith.. 60
19. **Catullus, Tibullus, Ovid, and Propertius,** Selections from, by W. M. Donne.......... 80
20. **Suetonius** and the later Latin Writers, Selections from, by W. M. Donne.......... 80
21. **The Satires of Juvenal,** by T. H. S. Escott, M.A., of Queen's College, Oxford.......... 60

GREEK SERIES.

1. **Greek Introductory Reader,** by H. Young........ $0 40
2. **Xenophon.** Anabasis, i. ii. iii., by H Young...... 40
3. **Xenophon.** Anabasis, iv. v. vi. vii., by H. Young... 40
4. **Lucian.** Select Dialogues, by T. H. L. Leary 40
5. **Homer.** Iliad, i. to vi., by T. H. L. Leary........... 60
6. **Homer.** Iliad, vii. to xii., by T. H. L. Leary........ 60
7. **Homer.** Iliad, xiii. to xviii., by T. H. L. Leary 60
8. **Homer.** Iliad, xix. to xxiv., by T. H. L. Leary 60
9. **Homer.** Odyssey, i. to vi., by T. H. L. Leary....... 60
10. **Homer.** Odyssey, vii. to xii., by T. H. L. Leary...... 60
11. **Homer.** Odyssey, xiii. to xviii., by T. H. L. Leary... 60
12. **Homer.** Odyssey, xix. to xxiv., and Hymns, by T. H. L. Leary.................................... 80
13. **Plato.** Apology, Crito, and Phædo, by J. Davies..... 80
14. **Herodotus,** i. ii., by T. H. L. Leary................ 60
15. **Herodotus,** iii. iv., by T. H. L. Leary............... 60
16. **Herodotus,** v. vi. vii., by T. H. L. Leary............ 60
17. **Herodotus,** viii. ix., and Index, by T. H. L. Leary.... 60
18. **Sophocles;** Œdipus Tyrranus, by H. Young......... 40
20. **Sophocles;** Antigone. by J. Milner................. 80
23. **Euripides;** Hecuba and Medea, by W. B. Smith...... 60
26. **Euripides;** Alcestis, by J. Milner.................. 40
30. **Aeschylus;** Prometheus Vinctus, by J. Davies....... 40
32. **Aeschylus;** Septem contra Thebas, by J. Davies..... 40
40. **Aristophanes;** Acharnians, by C. S. D. Townshend.. 60
41. **Thucydides,** i., by H. Young....................... 40
42. **Xenophon;** Agesilaus, by Le F. W. Jewett........ 60

www.ingramcontent.com/pod-product-compliance
Lightning Source LLC
LaVergne TN
LVHW021357110826
845150LV00007B/1690

* 9 7 8 1 4 2 5 5 1 3 7 7 1 *